Advances in Dye Degradation
(Volume 1)

Edited by

Paulpandian Muthu Mareeswaran
Department of Chemistry
College of Engineering
Anna University, Chennai - 600025
Tamil Nadu, India

&

Jegathalaprathaban Rajesh
Department of Chemistry
Saveetha School of Engineering
Saveetha Institute of Medical and Technical Sciences
Saveetha University
Chennai – 602 105
Tamil Nadu, India

Advances in Dye Degradation

(Volume 1)

Editors: Paulpandian Muthu Mareeswaran and Jegathalaprathaban Rajesh

ISBN (Online): 978-981-5179-54-5

ISBN (Print): 978-981-5179-55-2

ISBN (Paperback): 978-981-5179-56-9

Published by Bentham Science Publishers Pte. Ltd. Singapore. All Rights Reserved.

First published in 2023.

need for a court order if at any point you breach any terms of this License Agreement. In no event will any delay or failure by Bentham Science Publishers in enforcing your compliance with this License Agreement constitute a waiver of any of its rights.

3. You acknowledge that you have read this License Agreement, and agree to be bound by its terms and conditions. To the extent that any other terms and conditions presented on any website of Bentham Science Publishers conflict with, or are inconsistent with, the terms and conditions set out in this License Agreement, you acknowledge that the terms and conditions set out in this License Agreement shall prevail.

Bentham Science Publishers Pte. Ltd.
80 Robinson Road #02-00
Singapore 068898
Singapore
Email: subscriptions@benthamscience.net

BENTHAM SCIENCE

CONTENTS

FOREWORD

Synthetic dyes are a major part of dyes utilized by the textile industry. Organic dyes are the new chemicals that are extensively used in the textile industry nowadays. Most of these dyes are toxic and potentially carcinogenic in nature and pose major threats and create environmental problems. While dyeing, not all the dyes are adsorbed by the fabric materials and a considerable part is left as textile effluent. This effluent will mix with water bodies and create environmental hazards. Therefore, the treatment of textile effluents, especially dye removal, is an imperative step for environmental remediation. The removed dye is also a secondary pollutant to the environment. Therefore, the conversion of dye molecules to benign molecules is the ultimate step for environmental remediation. The conversion of carcinogenic dye molecules to an environmentally benign molecule is called dye degradation. From the industrial perspective, the process of conversion should be not only efficient but also cost-effective.

The editors have chosen the interesting aspects of dye degradation. The book "Advances in Dye Degradation" highlights the recent advantages of dye degradation pathways. However, without a basic understanding, the advanced perspective will have no meaning. Hence, Volume I of this book series deals with the fundamental aspects. In this work, the authors have discussed the nature of dye molecules, adverse effects, environmental problems caused by the dye molecules and their remediation process. The toxicity of the dyes is discussed in detail, and the application of nanotechnology, electrochemical and biological processes is also discussed. Since the advanced oxidation process is one of the important methods of the dye degradation process, the basic mechanism of photocatalytic dye degradation is also discussed in detail. This book deals with the basics of most of the relevant topics in the field of dye degradation, and it will certainly be useful for students as well as researchers.

C. Stella
Department of Oceanography and Coastal Area Studies
Thondi Campus Alagappa University
Karaikudi – 630 003
India

PREFACE

Dye degradation is an important step for the sustainable ecofriendly atmosphere in the dyeing and textile industry. The untreated textile effluent provides severe adverse effects on the ecosystem. This book deals with the mitigation of textile effluents. Chapter 1 is a fundamental chapter that describes the nature of dyes, classification, adverse effects and methods of removal of dyes from a bird's eye view. The aim of the first chapter is to give a basic idea of dyes and understand the necessity of textile effluent mitigation with respect to the ecosystem. This chapter also gives an overview of the most prevalent methods used for textile effluent mitigation and degradation processes.

Chapter 2 deals with the toxicity measurements of textile effluents. This chapter deals with the regulation of dyes, and analysis of dyes by means of various methods. This chapter provides a view of toxicological dosages like LD50 and LC50 values of dyes. Also, it provides details about the dose effect and dose response. This chapter also explains how the toxicity is evaluated using live animals. Chapter 3 deals with the microbial degradation of dyes as an overview. This chapter portrays microorganisms as an effective tool to mitigate textile effluents. The biological methods for decolorization and degradation of textile effluent are very successful and have various advantages over traditional procedures. Biological methods for removing toxic textile dyes are both environmentally friendly and cost-effective.

Chapter 4 pitches a platform about the utility of nanotechnology in dye degradation methods. Materials are always efficient, reusable and cost-effective over any other methods used in all the fields of technology. Nanoscience and technology provide effective solutions for various problems, and also have a deep impact in the field of dye degradation and textile effluent mitigation. The nanotechnology itself provides various methods for dye mitigation. This chapter provides an overview of the utilization of nanotechnology for textile effluent problems.

Chapter 5 deals with the electrochemical degradation of synthetic textile dyes from aqueous solution. Electrochemical methods are one of the effective methods for the treatment of effluent water. The CV, UV–Vis and chemical oxygen demand (COD) studies are the important parameters for degradation efficiency. This chapter deals with the process of applying electrochemical methods to textile companies for the mitigation of textile effluents. Chapter 6 also deals with the electrochemical processes of dye mitigation. This chapter concentrates on anodic oxidation processes. The mechanisms of electrochemical oxidation in anodic oxidation processes are explained in detail. The various anodic electrodes towards dye mitigation and degradation, their mechanism of action and efficiencies are reviewed.

Chapter 7 explains in detail the Z-scheme, which is a fundamental phenomenon for the photocatalytic degradation of dyes using various photoactive materials. Therefore, this Z-scheme has an impact on environmental remediation. The photocatalysts designed using Z-scheme have several advantages over the traditional photocatalytic processes, like efficient charge separation and electron transfer. Hence, it renders an efficient redox mechanism for the catalytic materials.

Chapter 8 deals with several photocatalytic materials for dye degradation, like metal oxides, metal sulfides, and metal ferrites. It also discusses with strategies to improve photocatalysts, such as doping the materials. This book strives to give collectively about the nature of dyes, the adverse effects of dyes and the basics of degradation methods. The editors aim to format the first volume of this series to pitch a basic idea of dye degradation. The upcoming volumes will develop the basic ideas into target-oriented dye mitigation for a better environment.

Paulpandian Muthu Mareeswaran
Department of Chemistry
College of Engineering
Anna University, Chennai - 600025
Tamil Nadu, India

&

Jegathalaprathaban Rajesh
Department of Chemistry
Saveetha School of Engineering
Saveetha Institute of Medical and Technical Sciences
Saveetha University
Chennai – 602 105
Tamil Nadu, India

List of Contributors

Arumugam Girija	Department of Chemistry, Velumanoharan Arts & Science College for Women, Ramanathapuram-623504, Tamil Nadu, India
Bosco Christin Maria Arputham Ashwin	Department of Chemistry, Pioneer Kumarasamy College, Nagarcoil-629003, Tamil Nadu, India
C. Christopher	Department of Chemistry, St.Xavier's College, Palayamkottai, Tirunelveli-627 002, India
Dhanesh Tiwary	Department of Chemistry, Indian Institute of Technology (Banaras Hindu University), Varanasi, 221005, India
G. Ravi	Department of Physics, Alagappa University, Karaikkudi, Tamil Nadu, India
Jegathalaprathaban Rajesh	Department of Chemistry, Saveetha School of Engineering, Saveetha Institute of Medical and Technical Sciences, Saveetha University, Chennai, Tamil Nadu, 602 105, India
Jeyaraj Dhaveethu Raja	Department of Chemistry, The American College, Tallakkulam, Madurai, 625 002, Tamil Nadu, India
K. Govindan	Environmental System Laboratory, Department of Civil Engineering, Kyung Hee University (Global Campus), Giheung-Gu, Yongin-Si, Gyeonggi-Do, 16705, Republic of Korea
Karupannan Aravindh	Research Centre, SSN College of Engineering, Kalavakkam, Chennai-603110, Tamil Nadu, India
Kiruthiga Kandhasamy	Department of Chemistry, Vellammal College of Engineering and Technology, Madurai–625 009, Tamil Nadu, India
Murugesan Sankarganesh	Department of Chemistry, Saveetha School of Engineering, Saveetha Institute of Medical and Technical Sciences, Saveetha University, Chennai, Tamil Nadu, 602 105, India
Nagaraj Revathi	Department of Chemistry, Ramco Institute of Technology, Rajapalayam, Virudhunagar, 626 117, Tamil Nadu, India
Paulpandian Muthu Mareeswaran	Department of Chemistry, College of Engineering, Anna University, Chennai-600025, Tamil Nadu, India
Perumalsamy Ramasamy	Research Centre, SSN College of Engineering, Kalavakkam, Chennai-603110, Tamil Nadu, India
Poovan Shanmugavelan	Department of Chemistry, School of Sciences, Tamil Nadu Open University, Saidapet, Chennai-600 015, Tamil Nadu, India
R. Jagatheesan	Department of Chemistry, Vivekanandha College of Arts and Sciences for Women (Autonomous), Elayampalayam, Tiruchengode, Tamil Nadu-637 205, India
R. Yuvakkumar	Department of Physics, Alagappa University, Karaikkudi, Tamil Nadu, India

Sai Sathish Ramamurthy   STAR Laboratory, Central Research Instruments Facility (CRIF), Department of Chemistry, Sri Sathya Sai Institute of Higher Learning, Prasanthi Nilayam, Puttaparthi, Anantapur, Andhra Pradesh, India

Seemesh Bhaskar   STAR Laboratory, Central Research Instruments Facility (CRIF), Department of Chemistry, Sri Sathya Sai Institute of Higher Learning, Prasanthi Nilayam, Puttaparthi, Anantapur, Andhra Pradesh, India
Department of Chemistry, Indian Institute of Technology (IIT) Bombay, Powai, Mumbai-400076, Maharashtra, India

Sheeba Daniel   Department of Chemistry, Holy Cross College (Autonomous), Nagercoil–629 004, Tamil Nadu, India

SP. Keerthana   Department of Physics, Alagappa University, Karaikkudi, Tamil Nadu, India

Suresh Kumar Pandey   Department of Chemistry, Indian Institute of Technology (Banaras Hindu University), Varanasi, 221005, India

Venkatesan Sethuraman   Research and Development, New Energy Storage Technology, Lithium-Ion Battery Division, Amara Raja Batteries Ltd, Karakambadi, Andhra Pradesh-517 520, India

CHAPTER 1

Dye Degradation - Basics and Necessity

Kiruthiga Kandhasamy[1], Sheeba Daniel[2], Poovan Shanmugavelan[3] and **Paulpandian Muthu Mareeswaran[4,*]**

[1] *Department of Chemistry, Vellammal College of Engineering and Technology, Madurai-625 009, Tamil Nadu, India*

[2] *Department of Chemistry, Holy Cross College (Autonomous), Nagercoil-629 004, Tamil Nadu, India*

[3] *Department of Chemistry, School of Sciences, Tamil Nadu Open University, Saidapet, Chennai-600 015, Tamil Nadu, India*

[4] *Department of Chemistry, College of Engineering, Anna University, Chennai-600025, Tamil Nadu, India*

Abstract: Without colour, life is incomplete. Dye refers to the compounds that give goods their colour. Even though natural dyes have been used for generations, their limitations have led to the development of synthetic dyes. By addressing the history and significance of natural dyes, the limitations of natural dyes, the introduction of synthetic dyes, the negative effects of synthetic dyes, and an overview of several techniques used for the treatment of disposed dyes in the environment, this chapter serves as a foundation for the discussion of the entire upcoming book. The goal of this chapter is to provide a brief overview of the need for and the concept of dye degradation.

Keywords: Colour index, Degradation, Natural dyes, Oxidation, Synthetic dyes.

INTRODUCTION

Dyes are coloured substances that adhere to the substrate and give items their colour. Otto N. Witt developed a dyeing theory in 1876 that was based on functional groups like auxochrome and chromophore. According to his idea, certain auxochromic groups, which are responsible for dyeing properties, and certain unsaturated chromophoric groups, which are responsible for colour, are present in all coloured organic compounds (also known as chromogens) [1]. Dyes absorb visible wavelength ranges of radiation, and the appearance of colour depends on the wavelength ranges that are both absorbed and reflected. The term "visible" was created since the human eye can perceive light between 380 nm

** **Corresponding author Paulpandian Muthu Mareeswaran:** Department of Chemistry, College of Engineering, Anna University, Chennai-600025, Tamil Nadu, India; E-mail: muthumareeswaran@gmail.com*

Paulpandian Muthu Mareeswaran & Jegathalaprathaban Rajesh (Eds.)

(violet) and 700 nm (red) [2]. Two distinct indices are used to represent the commercial dyes. First The first is called a "colour index generic number" (CIGN), and it is used by businesses. The second one is the colour index constitution number (CICN), which has to do with the chemical makeup of the dye and is primarily employed by producers and academics [3].

Based on the chemical components found in the compounds, which determine the colour of dye, dyes are divided into numerous categories (Fig. **1**) [4].

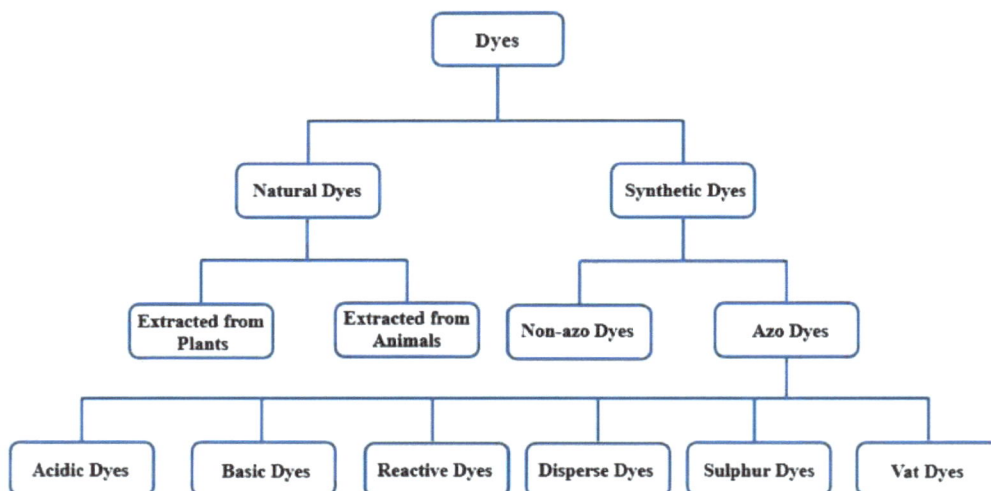

Fig. (1). Classification of dyes.

NATURAL DYES

Natural dyes and synthetic dyes are the two main classifications, which are based on their manufacturing techniques. Animals and other plant components, including the root, bark, leaf, flower, fruit, and seed, are used to make natural colours [5]. All civilizations have a very long history associated with the dyeing industry. Chinese dyeing techniques have been used for 5000 years [6]. The Ajantha cave paintings from the Ellora caves in India date to between 600 and 1000 CE, and those at Sittanavasal belong to the seventh century [7, 8]. These paintings are painted with vegetable oil colours, which have been around for more than a thousand years, on lime plaster. In order to categorize natural dyes, several criteria are taken into consideration, including their chemical makeup (anthracenes, carotenoids, xanthophylls, flavonoids, betacyanins, tannis, indigo, and chlorophyll dyes), their sources (animal and plant sources), their application techniques (direct dyes, acidic dyes, and basic dyes), and their colour [4]. The portions of plants from which the colour is derived are used to further categorize them. Based on colour, one of the most common classes is made. They are listed in the Colour Index based on their uses and chemical makeup of natural dyes. According on the application category they fall under, natural dyes have their own area in the Colour Index. Most red colour dyes are made from plant bark [9]. The predominant colour that can be derived from most plant parts is yellow.

The plant *Iindigofera* tinctoria is the source of the significant blue dye known as indigo (Fig. **2**). 6000 years ago, this dye was used for the first time in Peru [10]. Indigo uprising occurred in Bengal, a region of the Indian Subcontinent, as a result of the extensive cultivation of this dye during the colonial era [11]. *Lichens*, which are fungi-algae composite creatures, are some of the organisms used to colour clothes. They come in a variety of colours, including orange, red, pink, and yellow. Because these creatures must be grown in an environment free of pollution and impurities, industrial-scale production is not feasible [12].

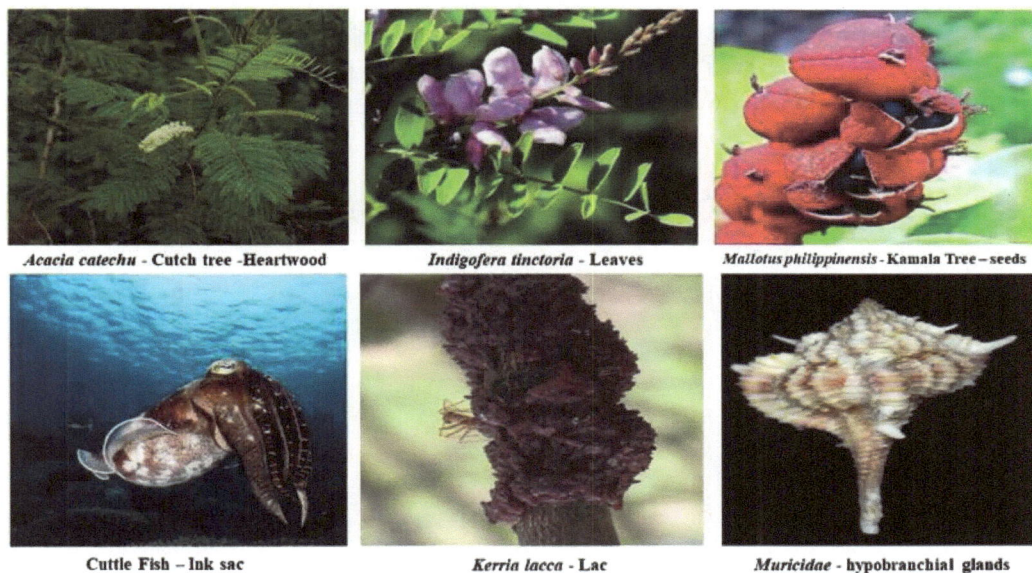

Acacia catechu - Cutch tree -Heartwood   Indigofera tinctoria - Leaves   Mallotus philippinensis - Kamala Tree – seeds

Cuttle Fish – Ink sac   Kerria lacca - Lac   Muricidae - hypobranchial glands

Fig. (2). Examples of natural colourants.

ADVANTAGES OF NATURAL DYES

Natural dyes are made from natural materials and are therefore not bad for the environment. Natural colours are biodegradable and renewable [2]. They can be used without risk, and there are no disposal issues. The plants used to make colours are frequently also used as medicines. Despite being used for generations, many of their therapeutic benefits have only recently come to light [12]. The dye made from henna, walnut, and alkanet, which is high in napthoquinone, also has antibacterial, antifungal, and anti-inflammatory properties. The use of natural dyes in dye-sensitive solar cells is one of their more recent applications. This is because a variety of dye compounds with plausible absorption mechanisms are readily available [13]. Natural dye-sensitized solar cells (NDSSC) have a high efficiency for converting solar energy [14]. For direct application of NDSSC, for instance, dye combinations like chlorophyll/anthocyanin and chlorophyll/betalain are used. The UV-Visible absorption spectra of natural dyes provide evidence of the cause. Anthocyanin dye absorbs light with a wavelength range of 480 to 580 nm. Chlorophyll dye absorbs light between the wavelengths of 500 and 600 nm, and by making synthetic alterations, this range can be increased to 600 to 700 nm. Betalain dye absorbs light between 470 and 600 nanometres (Fig. **3**). As a result, for effective energy harvesting, the absorption rage range spans most of the visible zone [15]. The energy conversion efficiency is improved by the physical coating or chemical anchoring of these dyes on semiconductor materials like TiO_2. Electrons can be injected into semiconductor materials by the stimulation of natural dye mixes with a wide range of absorption in the visible region [16]. Electron flow is created by the excited electrons from the dye that are injected to the materials (such ZnO, TiO_2, and Nb_2O_5). Increased surface area and effective dye mixture dispersion on the surface lead to efficient solar energy production [12].

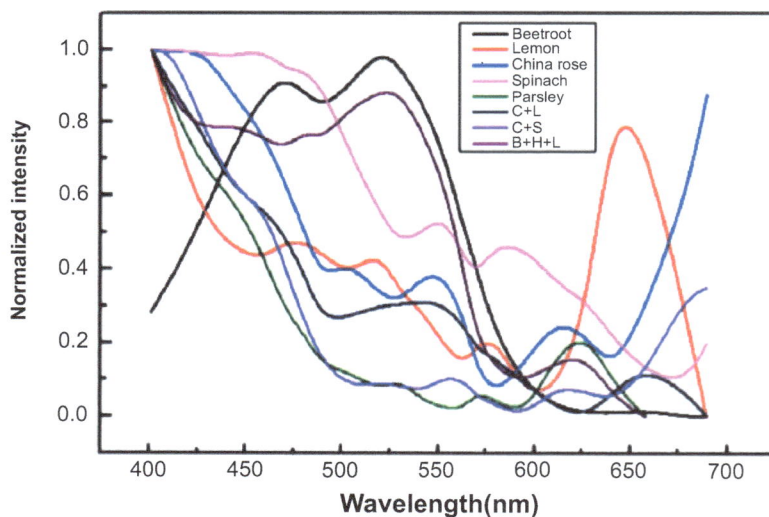

Fig. (3). Normalized absorption spectrum of a mixture of fresh natural dyes in ethanol solvent.

LIMITATIONS OF NATURAL DYES

In terms of commercial applications, natural dyes also have significant limitations. The main drawbacks of natural dyes are as follows [17 - 19].

Cost

As there is a large demand for dyes, there should be a high level of production. To produce dye, a significant number of raw materials are required, from which natural dye can be recovered. For the commercial manufacture of dyes, it is cost-effective to produce large quantities of the raw materials used to make dyes, such as the collection of leaves, flowers, bark, *etc*. Because of this, natural colours are pricey.

Colour

Natural dyes have a strong photobleaching of colour and low selectivity. Only selective dyes can be derived from natural dyes for selective colours. Synthetic dyes, on the other hand, come in a wider range.

Availability

Due to their cultivation circumstances, the availability of stating materials is limited. Consequently, it is challenging to produce all the basic ingredients in one location.

Harmful Effects

Natural dyes can sometimes be dangerous. Along with the environmentally harmful mordants, natural colours are employed. The use of organic solvents is part of the synthetic colour extraction process. Consequently, they also have a sizable number of environmental issues.

Sustainability

For the manufacturing of raw materials, land must be set aside for the cultivation of plants and trees that produce natural colours. In order to produce goods efficiently, this procedure needs labour.

SYNTHETIC DYES

The limitations of natural dyes prompted researchers to look for alternatives to meet demand and to begin developing synthetic colours. Dyes are artificial organic substances that are used in many industries, including textiles. Perkin created the first synthetic dye, known as mauveine, in 1856 [6]. The advantages of synthetic dyes over natural dyes are several, including their wide colour range, low cost, and resilience to fading from sunlight, water, and perspiration [19 - 21]. Natural dyes have been supplanted by synthetic dyes, and their industrial scale manufacture has increased dramatically—nearly 8×10^5 tons of synthetic dyes are produced annually. The textile sector uses over 75% of the world's dyestuffs. More than ten thousand different dyes and pigments are used in the global market to colour garments. The main users of synthetic dyes include also other sectors including printing, painting, and cosmetics. In addition to the heavy and light industrial sectors, one significant industry using synthetic dyes within acceptable limits is the food industry. The intriguing food colouring is utilized in jams, creams, and baking goods [22].

CLASSIFICATION OF DYES BASED ON APPLICATION

Direct Dyes

Direct dyes are colours that already have the ability to bond with fabric. Since most of them are water-soluble, direct dyes do not require mordants. The primary cause of the substantivity of direct dyes is the secondary valence bonding between dye and fabric. According to the Society of Dyers and Colourists (SDC), there are two categories for direct dyes: 1. Based on bleaching or leaving ability and 2. Chemical structure [23].

Reactive Dyes

Reactive dyes, a family of synthetic dyes that have had great success and outstanding fastness properties with most materials, are used in a variety of products. They have chromophores with pendant groups that can join with nucleophilic fibre material locations to generate covalent connections. The development of polyfunctional reactive dyes that can react or form bonds with two dye-fibre bonds is a breakthrough in reactive dyes [24 - 26].

Basic Dyes

Basic dyes are Cationic dyes. Most of them have a water-soluble nature and will adhere to the negative sites of the fibre materials. Due to electrostatic attraction, there is also a possibility of uneven dyeing. They are mostly used to dye fabrics made of wool, silk, and acrylic. Along with mordant, they are also used to color fabrics made of other materials, such cotton [27 - 29].

Acid Dyes

Acidic dyes are those that are frequently processed for dyeing under an acidic pH condition. The process through which these dyes interact with the fibres is crucial.

The ion exchange mechanism contributes to their strong binding. With the charged groups contained in the fibre materials, they are establishing ionic connections. The most popular fibre types for acid dying at low pH are protein fibres (such as wool and silk) and polyamides with amide groups [30 - 32].

Mordant or Chrome Dyes

Many dyes have a weak affinity for fabric. A mordant is a chemical that is used to increase the affinity of dyes for fabric. The word "modere" means "to bite" in Latin. Tannins, metallic mordants, and oil-mordants are the three different types of mordants. For stable coordination compounds with fibre materials, chrome dyes, which are acidic mordant dyes, can be used [33 - 37].

Disperse Dyes

These have a high substantivity to hydrophobic fibres, such as nylon, cellulose, cellulose acetate, and acrylic fibres, and are insoluble in water. Nonionic and water insolubleness are two crucial characteristics of dispersion dyes. Additionally, they are not altered chemically during the colouring process [38 - 40].

Vat Dyes

Vat dyes are insoluble in water. These dyes can be converted into a water-soluble form known as "*Leuco form*" by reducing them with inorganic ions. They were re-oxidized after the dyeing process to return them to their original state. The fabric receives stable colour as a result of the dyeing process. These dyes are insoluble in water due to their hefty conjugated structure. Additionally, the conjugation offers superior optical qualities that can be used in optoelectronic devices [41 - 43].

Sulphur Dyes

Sulphur colours are water-insoluble dyes that contain sulphur. They are significant in the dyeing business because of their inexpensive cost and affinity for cellulose. Alkali metals like sodium are used to decrease sulphur dyes to create water-soluble thiols. They are restored to their natural state after the colouring procedure [44].

Azoic Dyes

The most common synthetic dyes are azo dyes. Azo dyes make up 70% of synthetic organic dyes. They are easier to synthesize, have more structural variety, and high fastness. By diazotizing aromatic primary amine with amino and hydroxyl groups, azo dyes are synthesised. Azo dyes are categorized into monoazo, diazo, triazo, and polyazo dyes based on the number of azo links [45 - 47].

ADVANTAGES OF SYNTHETIC DYES

Crude oil (a fossil fuel) is the main source of synthetic dye because most synthetic dyes are made from petrochemicals. As a result, the raw resources are inexpensive and accessible. These dyes are effective at dying and produce a consistent, uniform colour. The availability of dyes for various types of materials is another crucial consideration. As opposed to natural dyes, synthetic dyes may be produced affordably and with less energy use by adjusting the hues and strength of the colours to suit the needs [19]. Since the colour is directly tied to the structural characteristics of dye molecules, synthetic dyes produce more striking coloration than conventional pigments. Whereas, the conventional pigments are mostly influenced by the physical properties [6]. For instance, dyes containing the azo, coumarin, and perylene groups are responsible for the vibrant colour [48]. Due to the commercial effect and the numerous benefits of synthetic dyes, two thirds of the synthetic dyes produced are commercial organic dyes with a wide variety of structural diversity and applications. Like the food sector and the pharmaceutical industry, synthetic dye-using industries are also growing. Safranine T, Thioflavin T, and other synthetic dyes make up most of the MRI contract agents and stains used for oncological investigation [49].

TOXIC EFFECT OF SYNTHETIC DYES

Synthetic dyes released into the environment, including water bodies, untreated or partially treated, have negative environmental effects (Fig. **4**). For dyeing, fixing, washing, and other processes, the textile industry uses enormous amounts of water, and 15% of the synthetic dyes used in these processes are released with the waste water [50, 51].

The effluents from textile factories contain a variety of organic and inorganic pollutants, including soaps, sequestering agents, dyes, pigments, chromium compounds, other heavy metals, chlorinated compounds, nitrates, sulphur, naphthol, formaldehyde, and benzidine [52]. Numerous harmful substances continue to exist in effluents even after the treatment process and cause multiple contaminants, such as soil, water, and air pollution [22]. The receiving water bodies (such as the sea, river, lake, natural ponds, and streams) are where the textile industry's effluents have the greatest impact on the living environment and ecosystem. Even at low dye concentrations (>1 mg/L), dyes can produce intense colours, but wastewater effluents typically have dye concentrations of 300 mg/L or higher along with other harmful substances. The pH impact of these effluents is their initial effect. The pH changes will cause a mass extinction of the animals and plants (planktons) present in the water bodies. The dark colour caused by the dyes prevents sunlight from penetrating, which significantly restricts the photosynthetic activity. As a result, the level of dissolved oxygen decreases [22].

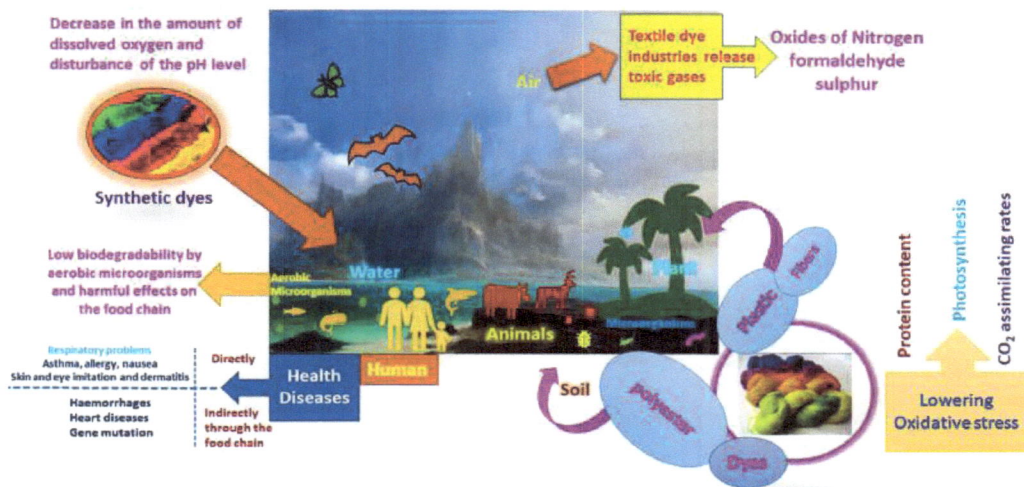

Fig. (4). The direct and indirect effect of synthetic dye on the environment [22].

DYE DEGRADATION TECHNIQUES

Researchers have consequently paid a lot of attention to the degradation of colours from textile industry effluents (Fig. **5**). Four categories—physical processes, chemical processes, biological processes, and combinatorial processes—are used to categorize the colour degradation approaches [53]. In addition to coagulation, reverse osmosis, photodegradation, ion exchange, oxidation, biodegradation, nanotechnology, an improved oxidation process, and adsorption, these procedures also include other elements [54]. The focus of current research is on safe, efficient, and environmentally acceptable methods for removing colours from contaminated water [55 - 57]. Microorganisms are likely to take a while for the chemical components of the dyes to biodegrade. All the traditional physical and chemical techniques used to eliminate dyes are frequently too expensive, only partially effective, and generate waste that is challenging to dispose of [58]. The majority of traditional techniques only successfully transfer dyes from one phase of water to another, which results in secondary contamination. Additional treatment is needed for the secondary contamination, and the procedure is not economical [59].

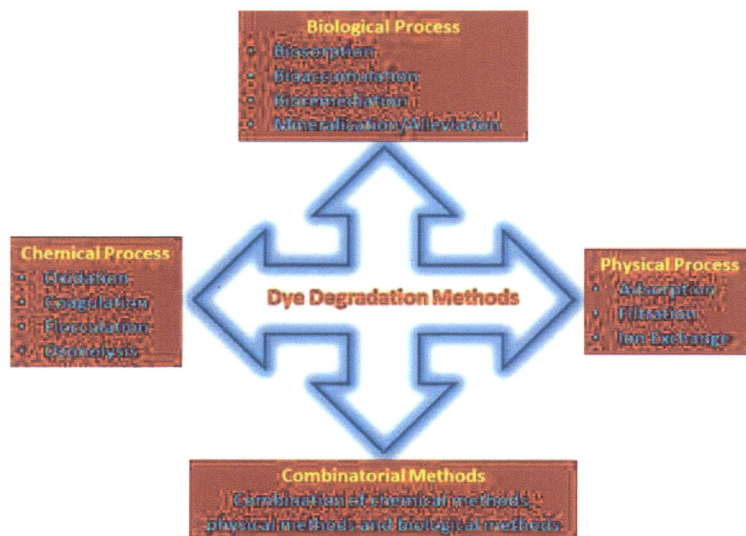

Fig. (5). Methods for degradation of dyes.

PHYSICAL PROCESSES

Physical dye removal is the removal of dyes without chemical modification. There are three types of physical processes; 1. Adsorption, 2. Filtration and 3. Ion exchange [60].

Adsorption

The solid-state removal of colours and pigments is centuries old technique [28]. Although the adsorption mechanism was not fully known in the past, adsorption technology has been studied since the turn of the twentieth century [61]. In many different industries, water-soluble synthetic dyes are utilized; these colours are typical industrial effluent water pollutants [62]. Due to their structural makeup, most synthetic dyes are stable when exposed to light and heat as well as resistant to aerobic digestion and oxidizing chemicals [63]. Adsorption is the effective method for removing dyes before they are broken down into benign chemicals [64]. Activated charcoal, silica, zeolites, bone charcoal, alumina, carbon molecular sieves, polymeric polymers, carbonized materials, and other materials are all accessible as adsorbents (Fig. **6**) [65]. The efficiency and selectivity of the adsorption materials can be adjusted by physical activation and chemical surface modification [60]. There are two types of adsorption processes: physical and chemical. An essential technique for activating surface adsorption is carbonization. The base materials for this approach range from biological wastes to synthetic polymeric materials [66, 67].

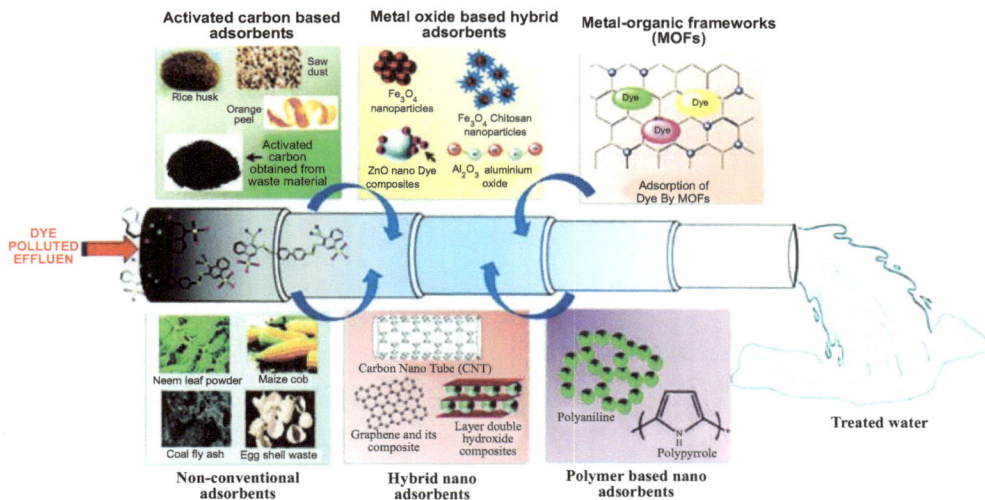

Fig. (6). Various adsorbent materials and the processes of adsorption [65].

Filtration

Adsorbent materials are always used in conjunction with filtering techniques [68]. Coagulation and filtration are frequently combined. Using filters such as filter papers, membrane filters, *etc.*, the suspended particles can be separated [69]. Adsorbent materials will be crucial if there is no coagulation in the effluents. By combining adsorbent and filter units into a single unit, porous materials and membrane technology make filtration a sophisticated process [70]. When treating waste water, the pore size becomes a determining factor in the filtration process (Fig. **7**) [71].

Ion Exchange

Cationic, anionic, and non-ionic dyes are the three main categories of dyes. Ionic dyes are unsightly pollutants that are challenging to remove. This occurs as a result of their electrical charge and water solubility. Since anionic dyes are acidic in nature and produce intense colours, they significantly modify pH. These dyes have a high level of ecosystem interference and are resistant to biological and chemical means of degradation. Adsorption is thought to be an effective strategy since charged molecules have a high inclination for it due to their individual charges and Vander Walls forces.

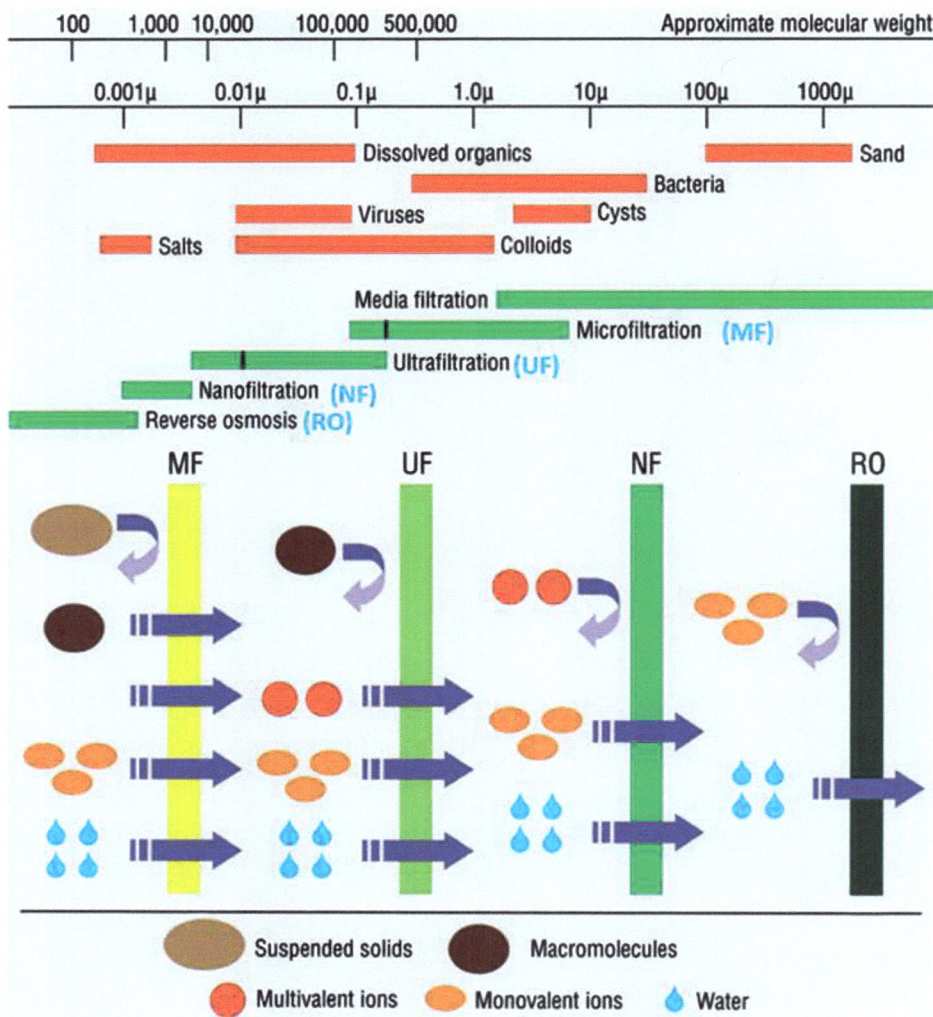

Fig. (7). Molecular weight cut off/pore size and component separation at various filtration level.

Ion exchangers are more effective at removing charged dyes. By releasing harmless ions and releasing the dye pollutants in the regenerant solution, ion exchangers are temporarily replacing the charged dyes [72]. Although the fundamental structure of resins is universal, they can be synthetically altered to match the needs of a certain application [73, 74]. The creation of macro reticular and microporous resin structures has improved dye removal capabilities (Fig. **8**) [75].

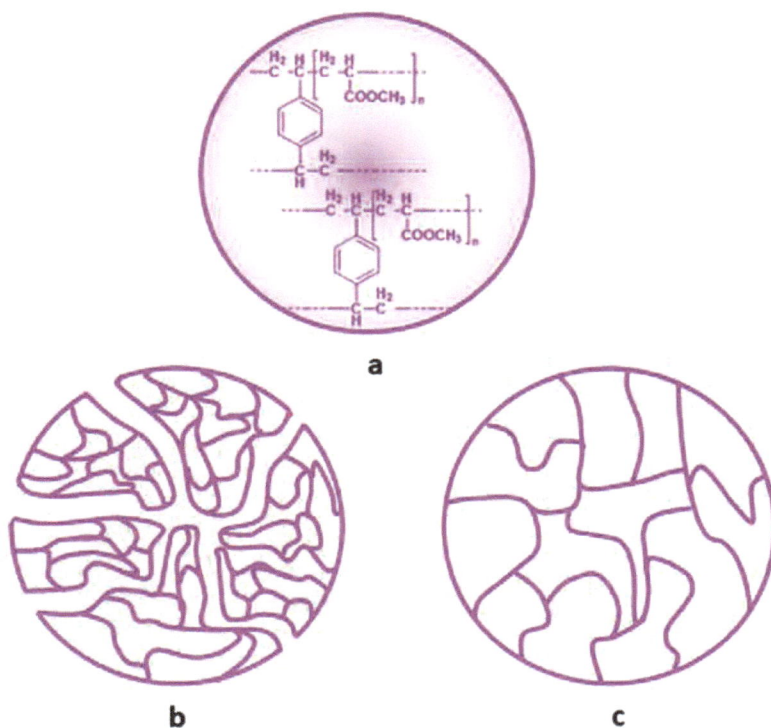

Fig. (8). Composition of resin matrices: (**a**) acrylic-divinylbenzene skeleton, (**b**) macroporous, (**c**) gel.

CHEMICAL PROCESSES

The chemical processes involved in dye degradation include a variety of oxidation, coagulation, and ozonolysis processes (Fig. **9**). One significant method of degradation among these processes is the oxidation of dye molecules into a benign product [76]. A variety of oxidants, including hydrogen peroxide, can be used to accomplish this. In order to achieve high efficiency, direct oxidation of dyes needs extreme circumstances such high temperature and pressure, which raises the overall cost of the degradation process. Due to their advantageous conditions and practical benefits, sophisticated oxidation techniques such catalysis, photocatalysis, electrocatalysis, ozonolysis, and Fenton processes are employed [77].

Advanced Oxidation

One of the most effective ways to degrade dyes into ecologically safe substances is using the advanced oxidation process (AOP) [50, 78]. Utilizing catalysis is the greatest technique to reduce energy consumption and costs for performing AOP under milder circumstances, which is necessary to accomplish efficient AOP [79]. The crucial stage in the reaction to break down the resistant and stable pollutant is the production of strong oxidizing radicals brought on by catalysis [80]. The key elements that influence the dye degradation efficiency are the radical production efficiency, reaction phase, electron transfer, selectivity, stability, surface area, and porosity. Small molecules that are friendly to the environment should be the result of the oxidation of hazardous pigments. As a result, catalysis depends greatly on the selectivity and sensitivity of the catalyst. In the end, the cost of catalyst manufacture and its capacity for reuse are crucial for industrial economic concerns [81].

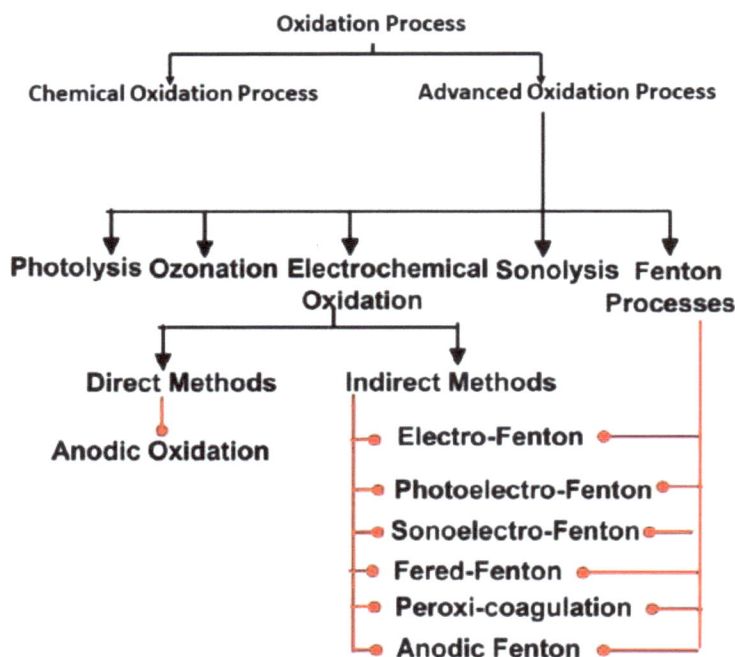

Fig. (9). Various oxidation processes of dye.

Photolysis

Dye degradation through photolysis involves the use of radiation (Fig. **10**). No radiation sensitizer is used in this situation [83]. Hydrogen peroxide is a typical oxidant utilized in this procedure. When exposed to UV light, the resulting •OH radical radicalizes the dye molecules, causing degradation [84]. The absence of any sensitizer in photolytic degradation is the primary distinction between photolysis and photocatalysis [77, 85 - 87].

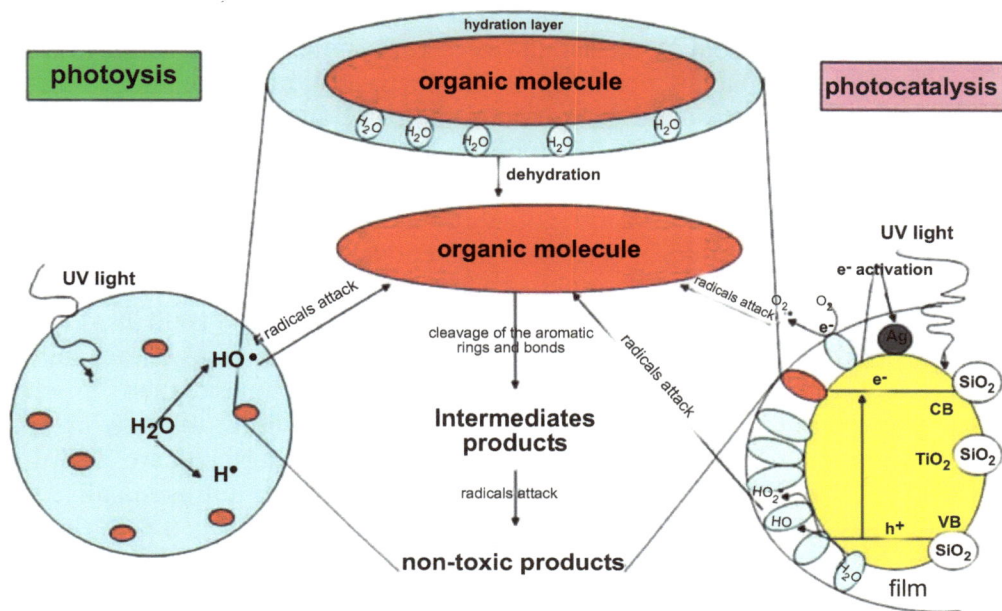

Fig. (10). Photolytic and photocatalytic dye degradation mechanism [82].

Photocatalysis

A sensitizer is employed during the photocatalysis (Fig. **10**). The two main types of photocatalysis are heterogeneous catalysis and homogeneous catalysis [88]. Heterogeneous catalysis has shown to be an effective treatment approach for managing hazardous compounds like dyes [89]. The ability to oxidize low concentration dyes at ppb levels is one of the benefits of photocatalysis [90]. The photooxidation process is an efficient method of treating effluents since it is a fully active, inexpensive catalyst and reactor system. Since light is used as the energy source, secondary risks like spent catalyst and secondary pollution like thermal pollution are avoided in this process [91 - 94].

Sonolysis

Chemicals can be broken down *via* sonolysis, which involves ultrasound (US). This energy source is completely distinct from conventional energy sources like heat, light, or ionizing radiation. Utilizing a US with a frequency range of 20-1000 kHz, organic compounds are broken down into small molecules. Because of this, this method is also effective for treating industrial effluents [95]. Since sound is a pure source of energy like light and there are no byproducts from using US, environmental sonochemistry is a rapidly expanding field [96]. Simple techniques for applying ultrasonic irradiation include submerging the reactor in an ultrasonic bath or placing the ultrasonic generator directly within the treatment facility [97]. The cheapest way to produce ultrasound is in ultrasonic cleaning bath, however the sonochemical impact it produces is ineffective at breaking down organic chemicals [98]. The ultrasound generated in the solution during sonolysis causes the dissolved gases to erupt into tiny bubbles. These micro-bubbles expand while the sonication continues, and at one point they collapse adiabatically by releasing high warmth in a little period of time. Cavitation is the development, expansion, and abrupt collapse of micro-bubbles that cause localized, transient "hot spots" in an irradiated liquid [99]. The thermal dissociation of water molecules into •H and •OH was caused by high temperatures (5000 K) and pressures (1000 atm) caused by cavitation in an aqueous solution. The interior of a collapsing gaseous bubble, the interface between the gas bubble and bulk liquid, where there are high temperature gradients (roughly 1000-2000 K), and the bulk solution at room temperature, which receives diffused •OH from the interface, are the three potential reaction sites in homogeneous liquids exposed ultrasonically. The sonochemical effect, or the oxidation of organic molecules, occurs at the gas-liquid interface because of the presence of •OH as well as, to a lesser extent, in the bulk solution because of diffused •OH. Along with oxidation *via* •OH, dyes also undergo direct pyrolytic breakdown. Using •OH, non-volatile substances like hydrophilic and dye molecules are broken down both in bulk solution and at the interface. On the other hand, pyrolytic breakdown occurs inside the bubbles to disintegrate the hydrophobic molecules (Fig. **11**) [100].

Electrochemical Dye Degradation Process

The electrochemical advanced oxidation processes (EAOPs), among other AOPs, attracted attention because of their potential applications [101]. When an electric current and a catalyst are applied, a large amount of hydroxyl radicals is produced, which destroy the dye molecules in the EAOP [102]. Using electrochemical technologies in remote locations offers benefits like automation and control. Electrochemically induced coagulation (EC), electrochemical reduction, and electrochemical oxidation are the three main electrochemical processes [103]. When Al or Fe electrodes are used as sacrifices during electrocoagulation, metal ions are released into the treatment solution at the application current. Inducing coagulations are these metal ions. Internal micro-electrolysis is one type of electrocoagulation [104 - 106]. Microelectrolysis performs galvanic cell-like functions. When treatment effluents are combined with iron chips and granular activated carbon, the interaction between the iron and activated carbon causes the treatment effluent in its whole to release electrons. The electron of mixture flow has the ability to break down organic contaminants like dyes into tiny benign molecules [102].

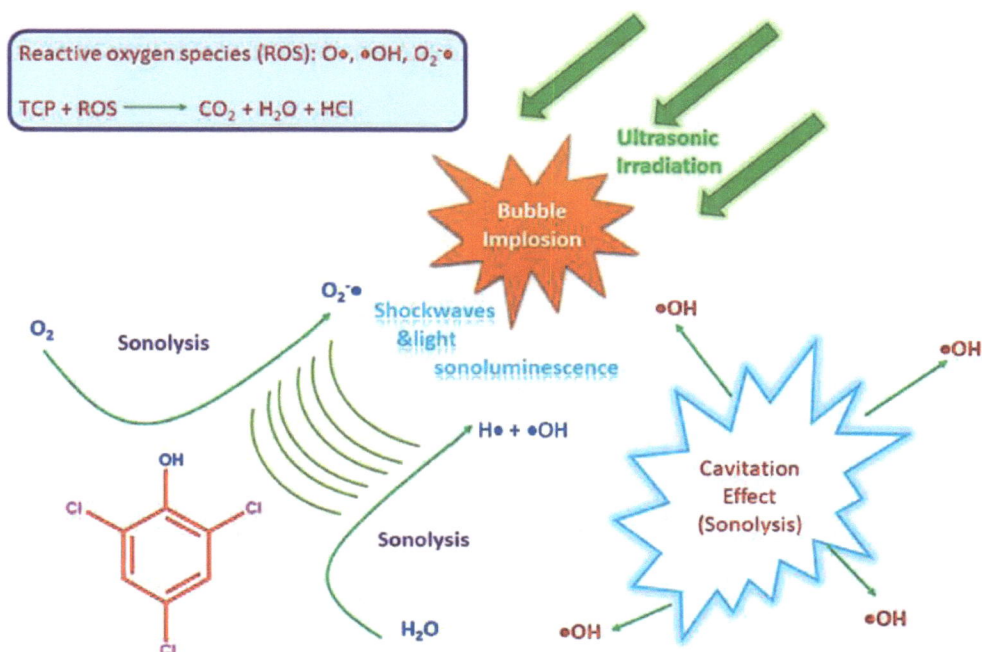

Fig. (11). Sonolysis mechanism of dye degradation [100].

Organic dyes are reduced at the cathode using electrochemical reduction to degrade them. However, electrochemical performance is less effective than oxidation because of side processes including hydrogen evolution [107]. Direct oxidation and mediated oxidation are the two fundamental electrooxidation processes. Dye is directly oxidized at the anode surface during the direct oxidation process. Even though the anode surface appears to have a high efficiency, the reaction is still possible. As a result, the amount of oxidation that may occur in a given amount of time limits its use in industry. The process of mediating oxidation entails the creation of active species that have the ability to oxidize the dye molecules found in the bulk solution. Reagents such as catalyst are added to the bulk solution to produce reactive species. The catalysts play a role in the creation of reactive species like singlet oxygen and occasionally they also work to lower the energy required for reactive species activation, which makes it feasible to degrade dyes more efficiently. It is also important to note that, in addition to the mediated oxidation process, direct oxidation also takes place at the electrode surface (Fig. **12**) [108].

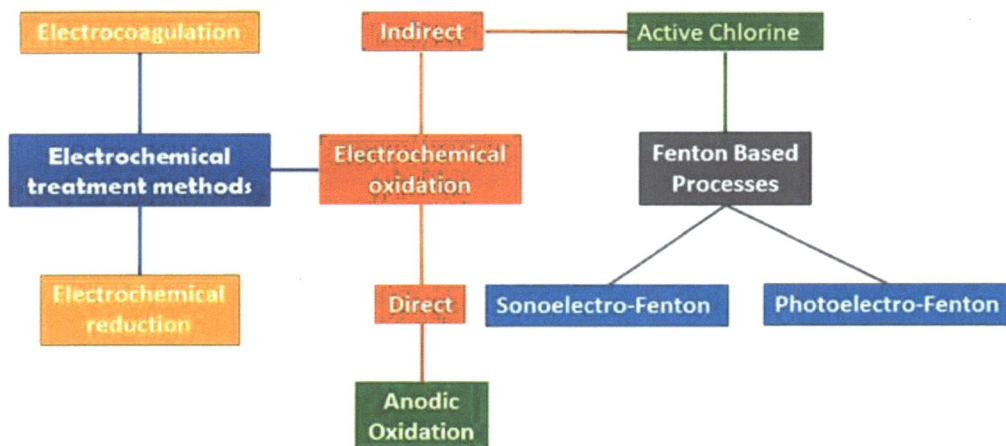

Fig. (12). General classification of electrochemical treatment processes.

Fenton Process

H.J.H. Fenton developed the Fenton reaction in 1894 and observed that ferrous (Fe^{2+}) salts may activate H_2O_2 to oxidize tartaric acid [109]. The Fenton (F) process is the name given to the reactive species that are produced when hydrogen peroxides interact with iron ions. Utilizing the reactive species created, organic pollution compounds like dyes are oxidized (Fig. **13**). Hydrogen peroxide's interaction with ferrous ions causes the ferric ion to be oxidized, which produces a hydroxyl radical. The dye molecules in the effluents are attacked by the hydroxyl radicals, which break them down into smaller, harmless molecules [110]. The Fenton procedure has benefits including handling under pressure and at room temperature. The chemicals are inexpensive, and handling and storing them are relatively simple. The Fenton process can be used with other methods to create effective dye degradation, and it can be tailored to the specific needs of the industry [111 - 117].

Ozonolysis

Ozonolysis is the process of oxidizing organic contaminants like colours utilizing ozone (O_3) as an oxidant [118]. One of the main justifications for using ozone as an oxidant is its high oxidation potential (2.08 V). Due to the large oxidation potential, oxidation at low concentrations is also achievable in this approach and is far more efficient than with other oxidants. Ozonolysis is a safe process in which no harmful substances are produced as byproducts [119]. For ozonolysis, there are two distinct pathways: direct oxidation and radical reaction. In the direct oxidation process, the ozone molecule interacts with organic molecules through functional groups like double bonds and aromatic groups. Most of this process takes place when the pH is acidic [120]. Ozone-produced hydroxide radicals react with organic substances at alkaline pH, which causes deterioration. Due to the characteristics of the hydroxyl radical, the interaction of radical processes is generally non-specific [121]. Organic molecules can be oxidized to form organic acids, which can locally change the pH. Through this procedure, dye molecules are directly oxidized by ozone [122].

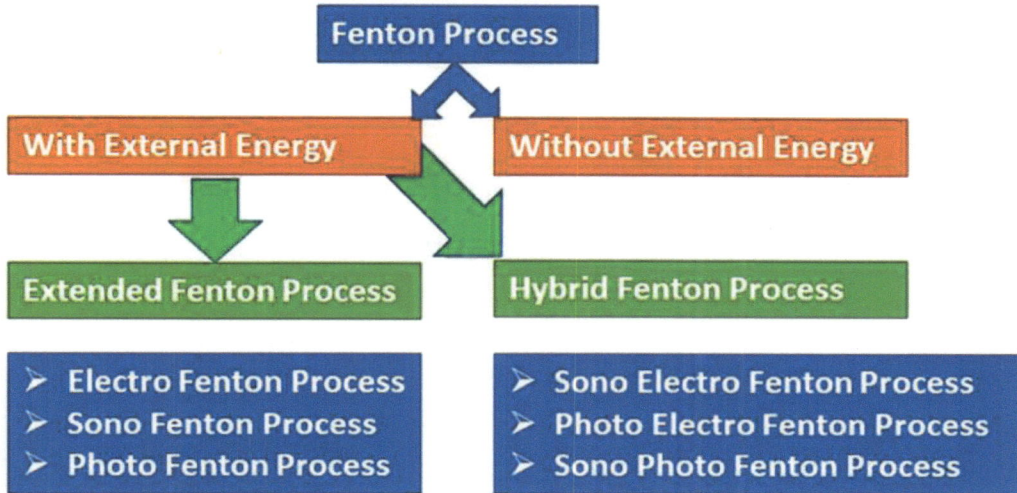

Fenton Process

With External Energy   **Without External Energy**

Extended Fenton Process   **Hybrid Fenton Process**

➤ **Electro Fenton Process**   ➤ **Sono Electro Fenton Process**
➤ **Sono Fenton Process**   ➤ **Photo Electro Fenton Process**
➤ **Photo Fenton Process**   ➤ **Sono Photo Fenton Process**

Fig. (13). Types of fenton processes.

BIOLOGICAL PROCESSES

Biosorption

Biosorption is the use of biological waste to absorb contaminants like dyes. For biosorption, bioproducts such agricultural wastes, industrial by-products, microorganisms, seeds, leaves, and bark of different plants are employed (Fig. **14**) [123]. Although biosorption techniques have been used for the treatment of effluents for more than three decades, the field of biosorption research is still advancing, as seen by the number of publications in recent years [124]. For the removal of dye, sophisticated biosorbent materials including biopolymeric composites and biopolymers are employed [125]. The choice of biosorbents is influenced by the physical and chemical properties of dye molecules [126]. The source of the biomass and the price are crucial considerations for industrial scale applications [127]. Among the several additional biosorbent materials, the dead biomass should be given priority [128].

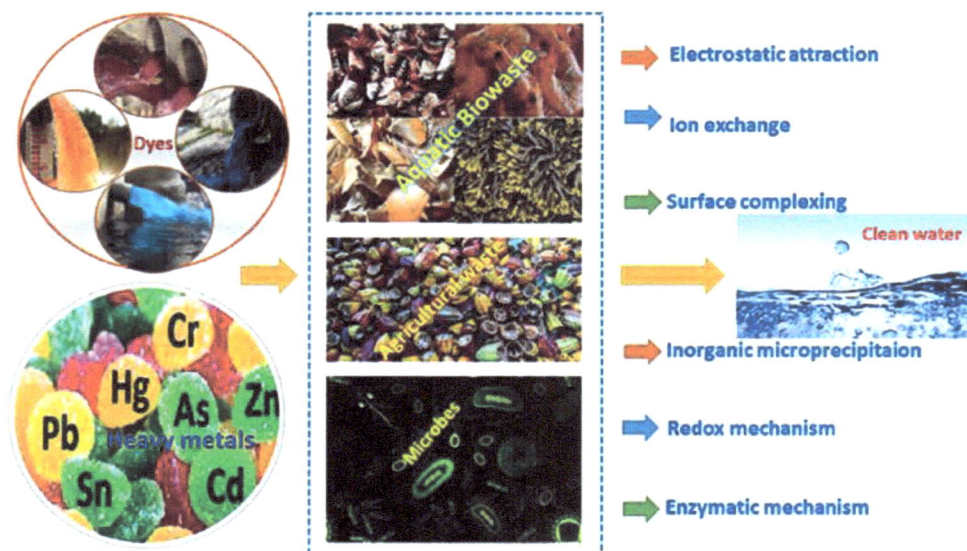

Fig. (14). Biomass sorbent materials and their mode of actions.

Bioaccumulation and Bioremediation

The definition of bioaccumulation is the intracellular accumulation of living sorbate. The first stage of bioaccumulation is when contaminants adsorb on sorbent. The biosorption procedure is comparable to this. The second mechanism involves dye pollution entering the intracellular space. Cell metabolic activity is required for this complicated process. Bioaccumulation results from the culture of organisms with living sorbates present [125, 129 - 131].

The process of biologically breaking down dyes and other contaminants into safe molecules is known as bioremediation [132]. Numerous benefits of bioremediation include techniques that are economical and environmentally friendly (Fig. **15**). The chemical structure of textile dyes is degraded by bacteria, which results in mineralization or transformation [133]. The microbes use a variety of methods to digest the contaminants and use them as food sources. These contaminants are degraded by digestion. The key to efficient degradation is the choice of microbes for the specific effluent [134]. Since metabolism is the primary mechanism for dye oxidation, enzymes for this process enable effective oxidation. The use of particular enzymes for particular dyes was shown to be highly effective [135, 136]. Environmental factors play a role in microbial development, which results in efficient decomposition [137]. To speed up the bioremediation process, the ambient conditions can be adjusted to encourage greater microbial growth [138]. Bioremediation provides a number of possible benefits over other traditional wastewater treatment methods. The bioremediation technique is very economical and environmentally benign because it is an entirely natural process. Because this method requires little maintenance and may be used on a wide scale, it can be implemented in higher surface waste treatment facilities [139]. This method's implementation is relatively easy and requires little effort. The number of labours required to manipulate this is reduced. The full degradation of contaminants is ensured by this method [140]. The main drawbacks of this technology are that it uses biodegradable contaminants and requires skilled human labour. This process takes a lot of time as well [141]. The study of bioremediation is developing. In order to achieve complete breakdown of organic pollutants like dyes, efficient bioremediation involving new microbial development and environmental changes is being researched [142].

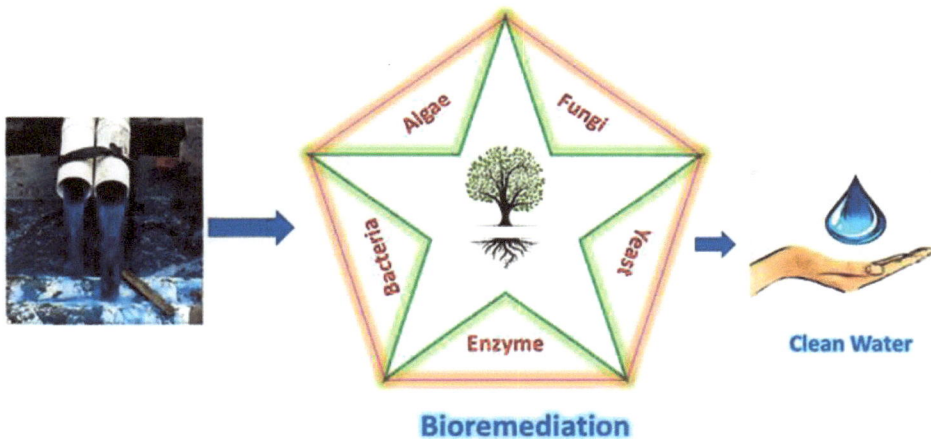

Fig. (15). Strategy of bioaccumulation and bioremidiation [132].

Mineralization and Alleviation

Mineralization is the process of turning dye-containing biomass that is both active and dead into harmless compounds (Fig. **16**) [143]. Elimination is the term for mineralization [144]. As mentioned in the previous sections, the mineralization process can be triggered by a number of factors, including chemical or biological processes [96].

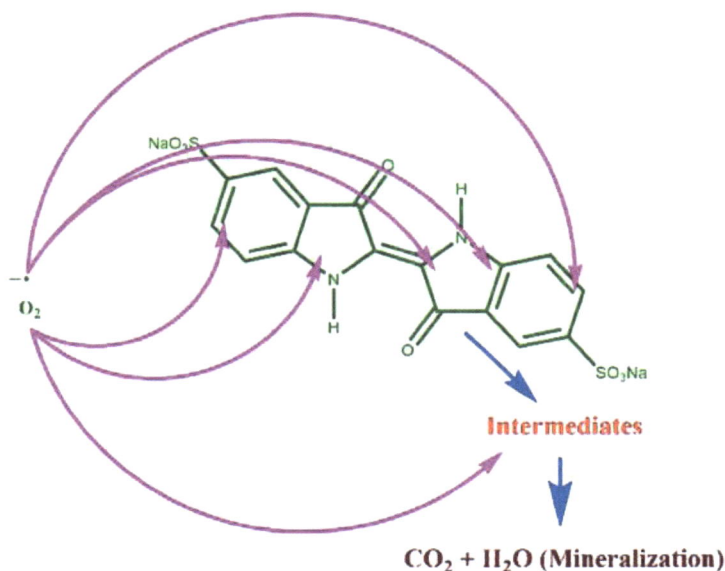

Fig. (16). Mineralization of azo dye by means of photocatalytic oxidation [145].

CONCLUSION

The dye molecule is a significant contaminant in textile wastewater. The dye molecules have a negative impact on the environment and human health. The structure of the dye molecules and the way they interact with light both make them cancer-causing substances. Both natural and artificial dye molecules have the characteristic of absorbing specific light wavelengths and sensitizing the surrounding environment through energy transfer and electron transfer reactions. Different degradation techniques are used to convert dyes into harmless chemicals. The field of dye degradation is affected equally by physical, chemical, and biological processes. The requirement and fundamental ideas of the most common dye degrading procedures were reviewed in this chapter. The individual concepts will be discussed in detail in the individual chapters.

REFERENCES

[1]   F. Habashi, "Witt and the theory of Dyeing", *Latest Trends Text. Fash. Design.,* vol. 3, pp. 640-643, 2019.

[2]   B.H. Patel, "Natural dyes", In: *Handbook of Textile and Industrial Dyeing.,* M. Clark, Ed., vol. Vol. 1. Woodhead Publishing, 2011, pp. 395-424.
        [http://dx.doi.org/10.1533/9780857093974.2.395]

[3]   Available from: http://colour-index.com /ci-explained

[4]   J.A. Kiernan, "Classification and naming of dyes, stains and fluorochromes", *Biotech. Histochem.,* vol. 76, no. 5-6, pp. 261-278, 2001.
        [http://dx.doi.org/10.1080/bih.76.5-6.261.278] [PMID: 11871748]

[5]   P.S. Vankar, "Chemistry of natural dyes", *Resonance,* vol. 5, no. 10, pp. 73-80, 2000.
        [http://dx.doi.org/10.1007/BF02836844]

[6]   L.D. Ardila-Leal, R.A. Poutou-Piñales, A.M. Pedroza-Rodríguez, and B.E. Quevedo-Hidalgo, "A brief
        history of colour, the environmental impact of synthetic dyes and removal by using laccases",
        Molecules, vol. 26, no. 13, p. 3813, 2021.
        [http://dx.doi.org/10.3390/molecules26133813] [PMID: 34206669]

[7]   Available from: https://whc.unesco.org/en/list/243/

[8]   Available from: https://www.tribuneindia.com/2005/20051127/spectrum/main3.htm

[9]   P. Gregory, "Classification of dyes by chemical structure", In: *The Chemistry and Application of
        Dyes.,* D.R. Waring, G. Hallas, Eds., Springer US: Boston, MA, 1990, pp. 17-47.
        [http://dx.doi.org/10.1007/978-1-4684-7715-3_2]

[10]   E.J.W. Barber, *Prehistoric Textiles.* Princeton University Press, 1991.
        [http://dx.doi.org/10.1515/9780691201412]

[11]   S. Bhattacharya, "The indigo revolt of bengal", *Soc. Sci.,* vol. 5, pp. 13-16, 1977.

[12]   U. Ahmed, and A. Anwar, "Application of natural dyes in dye-sensitized solar cells", In: *Dye-
        Sensitized Solar Cells.,* A.K. Pandey, S. Shahabuddin, M.S. Ahmad, Eds., Academic Press, 2022, pp.
        45-73.
        [http://dx.doi.org/10.1016/B978-0-12-818206-2.00008-6]

[13]   S. Sakshi, P.K. Singh, and V.K. Shukla, "Widening spectral range of absorption using natural dyes:
        Applications in dye sensitized solar cell", *Mater. Today Proc.,* vol. 49, pp. 3235-3238, 2022.
        [http://dx.doi.org/10.1016/j.matpr.2020.12.287]

[14]   G. Richhariya, A. Kumar, P. Tekasakul, and B. Gupta, "Natural dyes for dye sensitized solar cell: A
        review", *Renew. Sustain. Energy Rev.,* vol. 69, pp. 705-718, 2017.
        [http://dx.doi.org/10.1016/j.rser.2016.11.198]

[15]   H. Zhou, L. Wu, Y. Gao, and T. Ma, "Dye-sensitized solar cells using 20 natural dyes as sensitizers",
        J. Photochem. Photobiol. Chem., vol. 219, no. 2-3, pp. 188-194, 2011.
        [http://dx.doi.org/10.1016/j.jphotochem.2011.02.008]

[16]   S. Hao, J. Wu, Y. Huang, and J. Lin, "Natural dyes as photosensitizers for dye-sensitized solar cell",
        Sol. Energy, vol. 80, no. 2, pp. 209-214, 2006.
        [http://dx.doi.org/10.1016/j.solener.2005.05.009]

[17]   Available   from: https://textilevaluechain.in/news-insights/advantages-and-disadvantages-of-na-
        ural-dyes/

[18]   P.S. Vankar, *Natural Dyes for Textiles: Sources, Chemistry and Applications.* Elsevier Science, 2017.

[19]   A.K. Samanta, N. Awwad, and H.M. Algarni, *Chemistry and Technology of Natural and Synthetic
        Dyes and Pigments.* IntechOpen, 2020.
        [http://dx.doi.org/10.5772/intechopen.83199]

[20]   G. Ziarani, R. Moradi, N. Lashgari, and H.G. Kruger, *Metal-Free Synthetic Organic Dyes.* Elsevier
        Science, 2018.

[21]   S. Roy Maulik, C. Debnath, and P. Pandit, "Sustainable dyeing and printing of knitted fabric with
        natural dyes", In: *Advanced Knitting Technology.,* S. Maity, S. Rana, P. Pandit, K. Singha, Eds.,
        Woodhead Publishing, 2022, pp. 537-565.
        [http://dx.doi.org/10.1016/B978-0-323-85534-1.00008-8]

[22]   H.B. Slama, A. Chenari Bouket, Z. Pourhassan, F.N. Alenezi, A. Silini, H. Cherif-Silini, T. Oszako, L.
        Luptakova, P. Golińska, and L. Belbahri, "Diversity of synthetic dyes from textile industries, discharge
        impacts and treatment methods", *Appl. Sci.,* vol. 11, no. 14, p. 6255, 2021.
        [http://dx.doi.org/10.3390/app11146255]

[23]   N. Sekar, "12 - Direct dyes", In: *Handbook of Textile and Industrial Dyeing.,* M. Clark, Ed., vol. Vol. 1. Woodhead Publishing, 2011, pp. 425-445.
[http://dx.doi.org/10.1533/9780857093974.2.425]

[24]   A.H.M. Renfrew, "Reactive dyes for textile fibres: The chemistry of activated [pi]-bonds as reactive groups and miscellaneous topics", *Society of Dyers and Colourists,* 1999.

[25]   D.M. Lewis, "Developments in the chemistry of reactive dyes and their application processes", *Color. Technol.,* vol. 130, no. 6, pp. 382-412, 2014.
[http://dx.doi.org/10.1111/cote.12114]

[26]   D.M. Lewis, "The chemistry of reactive dyes and their application processes", In: *Handbook of Textile and Industrial Dyeing.,* M. Clark, Ed., vol. Vol. 1. Woodhead Publishing, 2011, pp. 303-364.
[http://dx.doi.org/10.1533/9780857093974.2.301]

[27]   N. Xu, R.L. Wang, D.P. Li, X. Meng, J.L. Mu, Z.Y. Zhou, and Z.M. Su, "A new triazine-based covalent organic polymer for efficient photodegradation of both acidic and basic dyes under visible light", *Dalton Trans.,* vol. 47, no. 12, pp. 4191-4197, 2018.
[http://dx.doi.org/10.1039/C8DT00148K] [PMID: 29479615]

[28]   G.M. Mohamed, W.E.E. Rashwan, T. El-Nabarawy, and A.N. El-Hendawy, "Removal of basic dyes from aqueous solution onto H2SO4-modified rice husk", *Egypt. J. Chem.,* vol. 64, no. 1, pp. 143-155, 2021.

[29]   M.I. Kiron, "Basic Dyes: Properties, classification, application, advantages and limitations", Available from: https://textilelearner.net/basic-dyes-properties-classification/

[30]   N. Sekar, "15 - Acid dyes", In: *Handbook of Textile and Industrial Dyeing.,* M. Clark, Ed., vol. Vol. 1. Woodhead Publishing, 2011, pp. 486-514.
[http://dx.doi.org/10.1533/9780857093974.2.486]

[31]   M.I. Kiron, "Characteristics, types and application of acid dyes", Available from: https://textilelearner.net/acid-dyes-properties/

[32]   S. Benkhaya, M. rabet, and El. Harfi, "A review on classifications, recent synthesis and applications of textile dyes", *Inorg. Chem. Commun.,* vol. 115, p. 107891, 2020.
[http://dx.doi.org/10.1016/j.inoche.2020.107891]

[33]   S. Har Bhajan, and K.A. Bharati, "4 - Mordants and their applications", In: *Handbook of Natural Dyes and Pigments.,* S. Har Bhajan, K.A. Bharati, Eds., Woodhead Publishing India, 2014, pp. 18-28.
[http://dx.doi.org/10.1016/B978-93-80308-54-8.50004-6]

[34]   Ö.E. İşmal, and L. Yıldırım, "Metal mordants and biomordants", In: *In The Impact and Prospects of Green Chemistry for Textile Technology,* ul, I. Shahid , B. S. Butola, Eds., Woodhead Publishing, 2019, pp. 57-82.

[35]   A.M. Thakker, and D. Sun, "Innovative plant-based mordants and colorants for application on cotton fabric", *J. Nat. Fibers,* vol. 19, no. 16, pp. 14346-14364, 2022.
[http://dx.doi.org/10.1080/15440478.2022.2064391]

[36]   A. Haji, "Dyeing of cotton fabric with natural dyes improved by mordants and plasma treatment progress in color", *Colorants Coatings,* vol. 12, no. 3, pp. 191-201, 2019.

[37]   G. Meier, "Application of chrome dyes to wool by the afterchrome process", *J. Soc. Dyers Colour.,* vol. 95, no. 7, pp. 252-257, 1979.
[http://dx.doi.org/10.1111/j.1478-4408.1979.tb03479.x]

[38]   M.L. Gulrajani, "Disperse dyes", In: *Handbook of Textile and Industrial Dyeing.,* M. Clark, Ed., vol. Vol. 1. Woodhead Publishing, 2011, pp. 365-394.
[http://dx.doi.org/10.1533/9780857093974.2.365]

[39]   A.D. Towns, "Developments in azo disperse dyes derived from heterocyclic diazo components", *Dyes Pigments,* vol. 42, no. 1, pp. 3-28, 1999.

[http://dx.doi.org/10.1016/S0143-7208(99)00005-4]

[40] A.M. Al-Etaibi, and M.A. El-Apasery, "A comprehensive review on the synthesis and versatile applications of biologically active pyridone-based disperse dyes", *Int. J. Environ. Res. Public Health,* vol. 17, no. 13, p. 4714, 2020.
[http://dx.doi.org/10.3390/ijerph17134714] [PMID: 32630027]

[41] J.F. Morin, "Recent advances in the chemistry of vat dyes for organic electronics", *J. Mater. Chem. C Mater. Opt. Electron. Devices,* vol. 5, no. 47, pp. 12298-12307, 2017.
[http://dx.doi.org/10.1039/C7TC03926C]

[42] R. Muthyala, *Chemistry and Applications of Leuco Dyes.* Springer US, 2006.

[43] M.R. Chetyrkina, F.S. Talalaev, L.V. Kameneva, S.V. Kostyuk, and P.A. Troshin, "Vat dyes: promising biocompatible organic semiconductors for wearable electronics applications", *J. Mater. Chem. C Mater. Opt. Electron. Devices,* vol. 10, no. 8, pp. 3224-3231, 2022.
[http://dx.doi.org/10.1039/D1TC04997F]

[44] K. Hunger, *Industrial Dyes: Chemistry, Properties, Applications.* Wiley-VCH, 2013.

[45] G.M. Ziarani, R. Moradi, N. Lashgari, and H.G. Kruger, "Chapter 4 - Azo Dyes", In: *In Metal-Free Synthetic Organic Dyes* Elsevier, 2018, pp. 47-93.

[46] A. Bafana, and D. Saravana, "ChakrabartiTapan, Azo dyes: past, present and the future", *Environ. Rev.,* vol. 19, no. NA, pp. 350-371, 2011.

[47] S. Benkhaya, S. M'rabet, and A. El Harfi, "Classifications, properties, recent synthesis and applications of azo dyes", *Heliyon,* vol. 6, no. 1, p. e03271, 2020.
[http://dx.doi.org/10.1016/j.heliyon.2020.e03271] [PMID: 32042981]

[48] S.J. Sharma, and N. Sekar, "Deep-red/NIR emitting coumarin derivatives - Synthesis, photophysical properties, and biological applications", *Dyes Pigments,* vol. 202, p. 110306, 2022.
[http://dx.doi.org/10.1016/j.dyepig.2022.110306]

[49] M. Urban, K. Durka, A. Kasprzak, T. Klis, A.P. Monkman, M. Piszcz, and K. Wozniak, "Excited-state photodynamics of pyrene-containing boronated dyes", *Dyes Pigments,* vol. 197, p. 109934, 2022.
[http://dx.doi.org/10.1016/j.dyepig.2021.109934]

[50] R. Javaid, and U.Y. Qazi, "Catalytic oxidation process for the degradation of synthetic dyes: An overview", *Int. J. Environ. Res. Public Health,* vol. 16, no. 11, p. 2066, 2019.
[http://dx.doi.org/10.3390/ijerph16112066] [PMID: 31212717]

[51] H. Yuan, L. Chen, Z. Cao, and F.F. Hong, "Enhanced decolourization efficiency of textile dye Reactive Blue 19 in a horizontal rotating reactor using strips of BNC-immobilized laccase: Optimization of conditions and comparison of decolourization efficiency", *Biochem. Eng. J.,* vol. 156, p. 107501, 2020.
[http://dx.doi.org/10.1016/j.bej.2020.107501]

[52] A. Srivastava, S. Shukla, N.K. Jangid, M. Srivastava, and R. Vishwakarma, "World of the Dye", In: *In Research Anthology on Emerging Techniques in Environmental Remediation, Management Association* IGI Global: Hershey: PA, USA, 2022, pp. 493-507.
[http://dx.doi.org/10.4018/978-1-6684-3714-8.ch026]

[53] V. Katheresan, J. Kansedo, and S.Y. Lau, "Efficiency of various recent wastewater dye removal methods: A review", *J. Environ. Chem. Eng.,* vol. 6, no. 4, pp. 4676-4697, 2018.
[http://dx.doi.org/10.1016/j.jece.2018.06.060]

[54] N.R.J. Hynes, J.S. Kumar, H. Kamyab, J.A.J. Sujana, O.A. Al-Khashman, Y. Kuslu, A. Ene, and B. Suresh Kumar, "Modern enabling techniques and adsorbents based dye removal with sustainability concerns in textile industrial sector -A comprehensive review", *J. Clean. Prod.,* vol. 272, p. 122636, 2020.
[http://dx.doi.org/10.1016/j.jclepro.2020.122636]

[55]   A. Nasar, and F. Mashkoor, "Application of polyaniline-based adsorbents for dye removal from water and wastewater—a review", *Environ. Sci. Pollut. Res. Int.*, vol. 26, no. 6, pp. 5333-5356, 2019.
[http://dx.doi.org/10.1007/s11356-018-3990-y] [PMID: 30612350]

[56]   W. Chen, J. Mo, X. Du, Z. Zhang, and W. Zhang, "Biomimetic dynamic membrane for aquatic dye removal", *Water Res.*, vol. 151, pp. 243-251, 2019.
[http://dx.doi.org/10.1016/j.watres.2018.11.078] [PMID: 30599283]

[57]   A. Maleki, M. Mohammad, Z. Emdadi, N. Asim, M. Azizi, and J. Safaei, "Adsorbent materials based on a geopolymer paste for dye removal from aqueous solutions", *Arab. J. Chem.*, vol. 13, no. 1, pp. 3017-3025, 2020.
[http://dx.doi.org/10.1016/j.arabjc.2018.08.011]

[58]   M.H. Munjur, M.N. Hasan, M.R. Awual, M.M. Islam, M.A. Shenashen, and J. Iqbal, "Biodegradable natural carbohydrate polymeric sustainable adsorbents for efficient toxic dye removal from wastewater", *J. Mol. Liq.*, vol. 319, p. 114356, 2020.
[http://dx.doi.org/10.1016/j.molliq.2020.114356]

[59]   P.T. Lum, K.Y. Foo, N.A. Zakaria, and P. Palaniandy, "Ash based nanocomposites for photocatalytic degradation of textile dye pollutants: A review", *Mater. Chem. Phys.*, vol. 241, p. 122405, 2020.
[http://dx.doi.org/10.1016/j.matchemphys.2019.122405]

[60]   L. Zhu, D. Shen, and K.H. Luo, "A critical review on VOCs adsorption by different porous materials: Species, mechanisms and modification methods", *J. Hazard. Mater.*, vol. 389, p. 122102, 2020.
[http://dx.doi.org/10.1016/j.jhazmat.2020.122102] [PMID: 32058893]

[61]   Y. Zhou, J. Lu, Y. Zhou, and Y. Liu, "Recent advances for dyes removal using novel adsorbents: A review", *Environ. Pollut.*, vol. 252, no. Pt A, pp. 352-365, 2019.
[http://dx.doi.org/10.1016/j.envpol.2019.05.072] [PMID: 31158664]

[62]   B.M. Thamer, A. Aldalbahi, M. Moydeen A, M. Rahaman, and M.H. El-Newehy, "Modified electrospun polymeric nanofibers and their nanocomposites as nanoadsorbents for toxic dye removal from contaminated waters: A Review", *Polymers*, vol. 13, no. 1, p. 20, 2020.
[http://dx.doi.org/10.3390/polym13010020] [PMID: 33374681]

[63]   J. Yun, Y. Wang, Z. Liu, Y. Li, H. Yang, and Z. Xu, "High efficient dye removal with hydrolyzed ethanolamine-Polyacrylonitrile UF membrane: Rejection of anionic dye and selective adsorption of cationic dye", *Chemosphere*, vol. 259, p. 127390, 2020.
[http://dx.doi.org/10.1016/j.chemosphere.2020.127390] [PMID: 32593817]

[64]   S. Madan, R. Shaw, S. Tiwari, and S.K. Tiwari, "Adsorption dynamics of Congo red dye removal using ZnO functionalized high silica zeolitic particles", *Appl. Surf. Sci.*, vol. 487, pp. 907-917, 2019.
[http://dx.doi.org/10.1016/j.apsusc.2019.04.273]

[65]   S. Dutta, B. Gupta, S.K. Srivastava, and A.K. Gupta, "Recent advances on the removal of dyes from wastewater using various adsorbents: A critical review", *Materials Advances*, vol. 2, no. 14, pp. 4497-4531, 2021.
[http://dx.doi.org/10.1039/D1MA00354B]

[66]   S. Sangon, A.J. Hunt, T.M. Attard, P. Mengchang, Y. Ngernyen, and N. Supanchaiyamat, "Valorisation of waste rice straw for the production of highly effective carbon based adsorbents for dyes removal", *J. Clean. Prod.*, vol. 172, pp. 1128-1139, 2018.
[http://dx.doi.org/10.1016/j.jclepro.2017.10.210]

[67]   Z. Heidarinejad, M.H. Dehghani, M. Heidari, G. Javedan, I. Ali, and M. Sillanpää, "Methods for preparation and activation of activated carbon: a review", *Environ. Chem. Lett.*, vol. 18, no. 2, pp. 393-415, 2020.
[http://dx.doi.org/10.1007/s10311-019-00955-0]

[68]   M.M.R. Feathers, *Membrane Technology: A Break Through in Water Treatment.* Water Conditioning and Purification International, 2009.

[69] P.S. Calabrò, S. Bilardi, and N. Moraci, "Advancements in the use of filtration materials for the removal of heavy metals from multicontaminated solutions", *Curr. Opin. Environ. Sci. Health,* vol. 20, p. 100241, 2021.
[http://dx.doi.org/10.1016/j.coesh.2021.100241]

[70] N. Grishkewich, N. Mohammed, S. Wei, M. Vasudev, Z. Shi, R.M. Berry, and K.C. Tam, "Dye removal using sustainable membrane adsorbents produced from melamine formaldehyde–cellulose nanocrystals and hard wood pulp", *Ind. Eng. Chem. Res.,* vol. 59, no. 47, pp. 20854-20865, 2020.
[http://dx.doi.org/10.1021/acs.iecr.0c04033]

[71] P.S. David, A. Karunanithi, and N.N. Fathima, "Improved filtration for dye removal using keratin–polyamide blend nanofibrous membranes", *Environ. Sci. Pollut. Res. Int.,* vol. 27, no. 36, pp. 45629-45638, 2020.
[http://dx.doi.org/10.1007/s11356-020-10491-y] [PMID: 32803596]

[72] J. Joseph, R.C. Radhakrishnan, J.K. Johnson, S.P. Joy, and J. Thomas, "Ion-exchange mediated removal of cationic dye-stuffs from water using ammonium phosphomolybdate", *Mater. Chem. Phys.,* vol. 242, p. 122488, 2020.
[http://dx.doi.org/10.1016/j.matchemphys.2019.122488]

[73] S. Karcher, A. Kornmüller, and M. Jekel, "Anion exchange resins for removal of reactive dyes from textile wastewaters", *Water Res.,* vol. 36, no. 19, pp. 4717-4724, 2002.
[http://dx.doi.org/10.1016/S0043-1354(02)00195-1] [PMID: 12448513]

[74] H. Peng, and J. Guo, "Removal of chromium from wastewater by membrane filtration, chemical precipitation, ion exchange, adsorption electrocoagulation, electrochemical reduction, electrodialysis, electrodeionization, photocatalysis and nanotechnology: A review", *Environ. Chem. Lett.,* vol. 18, no. 6, pp. 2055-2068, 2020.
[http://dx.doi.org/10.1007/s10311-020-01058-x]

[75] E. Kociołek-Balawejder, E. Stanisławska, I. Jacukowicz-Sobala, and M. Jasiorski, "Anomalous effect of Cu_2O and CuO deposit on the porosity of a macroreticular anion exchanger", *J. Nanopart. Res.,* vol. 23, no. 6, p. 126, 2021.
[http://dx.doi.org/10.1007/s11051-021-05246-w]

[76] D. Kanakaraju, B.D. Glass, and M. Oelgemöller, "Advanced oxidation process-mediated removal of pharmaceuticals from water: A review", *J. Environ. Manage.,* vol. 219, pp. 189-207, 2018.
[http://dx.doi.org/10.1016/j.jenvman.2018.04.103] [PMID: 29747102]

[77] S. Yang, P. Wang, X. Yang, G. Wei, W. Zhang, and L. Shan, "A novel advanced oxidation process to degrade organic pollutants in wastewater: Microwave-activated persulfate oxidation", *J. Environ. Sci. (China),* vol. 21, no. 9, pp. 1175-1180, 2009.
[http://dx.doi.org/10.1016/S1001-0742(08)62399-2] [PMID: 19999962]

[78] S. Hisaindee, M.A. Meetani, and M.A. Rauf, "Application of LC-MS to the analysis of advanced oxidation process (AOP) degradation of dye products and reaction mechanisms", *Trends Analyt. Chem.,* vol. 49, pp. 31-44, 2013.
[http://dx.doi.org/10.1016/j.trac.2013.03.011]

[79] Q. Yang, Y. Ma, F. Chen, F. Yao, J. Sun, S. Wang, K. Yi, L. Hou, X. Li, and D. Wang, "Recent advances in photo-activated sulfate radical-advanced oxidation process (SR-AOP) for refractory organic pollutants removal in water", *Chem. Eng. J.,* vol. 378, p. 122149, 2019.
[http://dx.doi.org/10.1016/j.cej.2019.122149]

[80] P. Verma, and S.K. Samanta, "Microwave-enhanced advanced oxidation processes for the degradation of dyes in water", *Environ. Chem. Lett.,* vol. 16, no. 3, pp. 969-1007, 2018.
[http://dx.doi.org/10.1007/s10311-018-0739-2]

[81] J. Yu, H. Feng, L. Tang, Y. Pang, G. Zeng, Y. Lu, H. Dong, J. Wang, Y. Liu, C. Feng, J. Wang, B. Peng, and S. Ye, "Metal-free carbon materials for persulfate-based advanced oxidation process: Microstructure, property and tailoring", *Prog. Mater. Sci.,* vol. 111, p. 100654, 2020.

[http://dx.doi.org/10.1016/j.pmatsci.2020.100654]

[82]   A. Peter, A. Mihaly-Cozmuta, C. Nicula, L. Mihaly-Cozmuta, A. Jastrzębska, A. Olszyna, and L. Baia, "UV light-assisted degradation of methyl orange, methylene blue, phenol, salicylic acid, and rhodamine b: Photolysis versus photocatalyis", *Water Air Soil Pollut.,* vol. 228, no. 1, p. 41, 2017. [http://dx.doi.org/10.1007/s11270-016-3226-z]

[83]   F. Al-Momani, E. Touraud, J.R. Degorce-Dumas, J. Roussy, and O. Thomas, "Biodegradability enhancement of textile dyes and textile wastewater by VUV photolysis", *J. Photochem. Photobiol. Chem.,* vol. 153, no. 1-3, pp. 191-197, 2002. [http://dx.doi.org/10.1016/S1010-6030(02)00298-8]

[84]   M.A. Rauf, N. Marzouki, and B.K. Körbahti, "Photolytic decolorization of Rose Bengal by UV/H2O2 and data optimization using response surface method", *J. Hazard. Mater.,* vol. 159, no. 2-3, pp. 602-609, 2008. [http://dx.doi.org/10.1016/j.jhazmat.2008.02.098] [PMID: 18395977]

[85]   H. Zhang, L. Wang, P. Dong, S. Mao, P. Mao, and G. Liu, "Photolysis of the BODIPY dye activated by pillar[5]arene", *RSC Advances,* vol. 11, no. 13, pp. 7454-7458, 2021. [http://dx.doi.org/10.1039/D0RA08611H] [PMID: 35423231]

[86]   O.V. Istomina, S.K. Evstropiev, E.V. Kolobkova, and A.O. Trofimov, "Photolysis of diazo dye in solutions and films containing zinc and silver oxides", *Opt. Spectrosc.,* vol. 124, no. 6, pp. 774-778, 2018. [http://dx.doi.org/10.1134/S0030400X18060097]

[87]   N.A. Volkova, S.K. Evstrop'ev, O.V. Istomina, and E.V. Kolobkova, "Photolysis of diazo dye in aqueous solutions of metal nitrates", *Opt. Spectrosc.,* vol. 124, no. 4, pp. 489-493, 2018. [http://dx.doi.org/10.1134/S0030400X18040197]

[88]   H. Anwer, A. Mahmood, J. Lee, K.H. Kim, J.W. Park, and A.C.K. Yip, "Photocatalysts for degradation of dyes in industrial effluents: Opportunities and challenges", *Nano Res.,* vol. 12, no. 5, pp. 955-972, 2019. [http://dx.doi.org/10.1007/s12274-019-2287-0]

[89]   S. Sarkar, N.T. Ponce, A. Banerjee, R. Bandopadhyay, S. Rajendran, and E. Lichtfouse, "Green polymeric nanomaterials for the photocatalytic degradation of dyes: a review", *Environ. Chem. Lett.,* vol. 18, no. 5, pp. 1569-1580, 2020. [http://dx.doi.org/10.1007/s10311-020-01021-w] [PMID: 32837482]

[90]   T.R. Waghmode, M.B. Kurade, R.T. Sapkal, C.H. Bhosale, B.H. Jeon, and S.P. Govindwar, "Sequential photocatalysis and biological treatment for the enhanced degradation of the persistent azo dye methyl red", *J. Hazard. Mater.,* vol. 371, pp. 115-122, 2019. [http://dx.doi.org/10.1016/j.jhazmat.2019.03.004] [PMID: 30849565]

[91]   A. Ajmal, I. Majeed, R.N. Malik, H. Idriss, and M.A. Nadeem, "Principles and mechanisms of photocatalytic dye degradation on TiO 2 based photocatalysts: A comparative overview", *RSC Advances,* vol. 4, no. 70, pp. 37003-37026, 2014. [http://dx.doi.org/10.1039/C4RA06658H]

[92]   B. Neppolian, H.C. Choi, S. Sakthivel, B. Arabindoo, and V. Murugesan, "Solar/UV-induced photocatalytic degradation of three commercial textile dyes", *J. Hazard. Mater.,* vol. 89, no. 2-3, pp. 303-317, 2002. [http://dx.doi.org/10.1016/S0304-3894(01)00329-6] [PMID: 11744213]

[93]   G. Liu, and J. Zhao, "Photocatalytic degradation of dye sulforhodamine B: A comparative study of photocatalysis with photosensitization", *New J. Chem.,* vol. 24, no. 6, pp. 411-417, 2000. [http://dx.doi.org/10.1039/b001573n]

[94]   M. Rochkind, S. Pasternak, and Y. Paz, "Using dyes for evaluating photocatalytic properties: A critical review", *Molecules,* vol. 20, no. 1, pp. 88-110, 2014. [http://dx.doi.org/10.3390/molecules20010088] [PMID: 25546623]

[95]   Z. Eren, "Ultrasound as a basic and auxiliary process for dye remediation: A review", *J. Environ. Manage.,* vol. 104, pp. 127-141, 2012.
[http://dx.doi.org/10.1016/j.jenvman.2012.03.028] [PMID: 22495014]

[96]   S. Vajnhandl, and A. Majcen Le Marechal, "Ultrasound in textile dyeing and the decolouration/mineralization of textile dyes", *Dyes Pigments,* vol. 65, no. 2, pp. 89-101, 2005.
[http://dx.doi.org/10.1016/j.dyepig.2004.06.012]

[97]   P. Chowdhury, and T. Viraraghavan, "Sonochemical degradation of chlorinated organic compounds, phenolic compounds and organic dyes – A review", *Sci. Total Environ.,* vol. 407, no. 8, pp. 2474-2492, 2009.
[http://dx.doi.org/10.1016/j.scitotenv.2008.12.031] [PMID: 19200588]

[98]   S. Merouani, O. Hamdaoui, F. Saoudi, and M. Chiha, "Sonochemical degradation of Rhodamine B in aqueous phase: Effects of additives", *Chem. Eng. J.,* vol. 158, no. 3, pp. 550-557, 2010.
[http://dx.doi.org/10.1016/j.cej.2010.01.048]

[99]   E.A. Serna-Galvis, J. Porras, and R.A. Torres-Palma, "A critical review on the sonochemical degradation of organic pollutants in urine, seawater, and mineral water", *Ultrason. Sonochem.,* vol. 82, p. 105861, 2022.
[http://dx.doi.org/10.1016/j.ultsonch.2021.105861] [PMID: 34902815]

[100]   S. Anandan, V. Kumar Ponnusamy, and M. Ashokkumar, "A review on hybrid techniques for the degradation of organic pollutants in aqueous environment", *Ultrason. Sonochem.,* vol. 67, p. 105130, 2020.
[http://dx.doi.org/10.1016/j.ultsonch.2020.105130] [PMID: 32315972]

[101]   S.O. Ganiyu, C.A. Martínez-Huitle, and M.A. Oturan, "Electrochemical advanced oxidation processes for wastewater treatment: Advances in formation and detection of reactive species and mechanisms", *Curr. Opin. Electrochem.,* vol. 27, p. 100678, 2021.
[http://dx.doi.org/10.1016/j.coelec.2020.100678]

[102]   F. Ghanbari, and C.A. Martínez-Huitle, "Electrochemical advanced oxidation processes coupled with peroxymonosulfate for the treatment of real washing machine effluent: A comparative study", *J. Electroanal. Chem.,* vol. 847, p. 113182, 2019.
[http://dx.doi.org/10.1016/j.jelechem.2019.05.064]

[103]   P.V. Nidheesh, M. Zhou, and M.A. Oturan, "An overview on the removal of synthetic dyes from water by electrochemical advanced oxidation processes", *Chemosphere,* vol. 197, pp. 210-227, 2018.
[http://dx.doi.org/10.1016/j.chemosphere.2017.12.195] [PMID: 29366952]

[104]   O. Ganzenko, D. Huguenot, E.D. van Hullebusch, G. Esposito, and M.A. Oturan, "Electrochemical advanced oxidation and biological processes for wastewater treatment: A review of the combined approaches", *Environ. Sci. Pollut. Res. Int.,* vol. 21, no. 14, pp. 8493-8524, 2014.
[http://dx.doi.org/10.1007/s11356-014-2770-6] [PMID: 24965093]

[105]   G. Chen, "Electrochemical technologies in wastewater treatment", *Separ. Purif. Tech.,* vol. 38, no. 1, pp. 11-41, 2004.
[http://dx.doi.org/10.1016/j.seppur.2003.10.006]

[106]   E. Butler, Y.T. Hung, R.Y.L. Yeh, and M. Suleiman Al Ahmad, "Electrocoagulation in wastewater treatment", *Water,* vol. 3, no. 2, pp. 495-525, 2011.
[http://dx.doi.org/10.3390/w3020495]

[107]   N.P. Shetti, S.J. Malode, R.S. Malladi, S.L. Nargund, S.S. Shukla, and T.M. Aminabhavi, "Electrochemical detection and degradation of textile dye Congo red at graphene oxide modified electrode", *Microchem. J.,* vol. 146, pp. 387-392, 2019.
[http://dx.doi.org/10.1016/j.microc.2019.01.033]

[108]   C.A. Martínez-Huitle, and M. Panizza, "Electrochemical oxidation of organic pollutants for wastewater treatment", *Curr. Opin. Electrochem.,* vol. 11, pp. 62-71, 2018.

[http://dx.doi.org/10.1016/j.coelec.2018.07.010]

[109]   A. Babuponnusami, and K. Muthukumar, "A review on Fenton and improvements to the Fenton process for wastewater treatment", *J. Environ. Chem. Eng.,* vol. 2, no. 1, pp. 557-572, 2014.
[http://dx.doi.org/10.1016/j.jece.2013.10.011]

[110]   M.E. Farshchi, H. Aghdasinia, and A. Khataee, "Modeling of heterogeneous Fenton process for dye degradation in a fluidized-bed reactor: Kinetics and mass transfer", *J. Clean. Prod.,* vol. 182, pp. 644-653, 2018.
[http://dx.doi.org/10.1016/j.jclepro.2018.01.225]

[111]   M. Zhang, H. Dong, L. Zhao, D. Wang, and D. Meng, "A review on Fenton process for organic wastewater treatment based on optimization perspective", *Sci. Total Environ.,* vol. 670, pp. 110-121, 2019.
[http://dx.doi.org/10.1016/j.scitotenv.2019.03.180] [PMID: 30903886]

[112]   V. Innocenzi, M. Prisciandaro, M. Centofanti, and F. Vegliò, "Comparison of performances of hydrodynamic cavitation in combined treatments based on hybrid induced advanced Fenton process for degradation of azo-dyes", *J. Environ. Chem. Eng.,* vol. 7, no. 3, p. 103171, 2019.
[http://dx.doi.org/10.1016/j.jece.2019.103171]

[113]   V. Poza-Nogueiras, E. Rosales, M. Pazos, and M.Á. Sanromán, "Current advances and trends in electro-Fenton process using heterogeneous catalysts – A review", *Chemosphere,* vol. 201, pp. 399-416, 2018.
[http://dx.doi.org/10.1016/j.chemosphere.2018.03.002] [PMID: 29529567]

[114]   T. Yu, and C.B. Breslin, "Graphene-modified composites and electrodes and their potential applications in the electro-fenton process", *Materials,* vol. 13, no. 10, p. 2254, 2020.
[http://dx.doi.org/10.3390/ma13102254] [PMID: 32422892]

[115]   E. Brillas, "A review on the photoelectro-Fenton process as efficient electrochemical advanced oxidation for wastewater remediation. Treatment with UV light, sunlight, and coupling with conventional and other photo-assisted advanced technologies", *Chemosphere,* vol. 250, p. 126198, 2020.
[http://dx.doi.org/10.1016/j.chemosphere.2020.126198] [PMID: 32105855]

[116]   P. Prete, A. Fiorentino, L. Rizzo, A. Proto, and R. Cucciniello, "Review of aminopolycarboxylic acids–based metal complexes application to water and wastewater treatment by (photo-)Fenton process at neutral pH", *Curr. Opin. Green Sustain. Chem.,* vol. 28, p. 100451, 2021.
[http://dx.doi.org/10.1016/j.cogsc.2021.100451]

[117]   M.M. Bello, A.A. Abdul Raman, and A. Asghar, "A review on approaches for addressing the limitations of Fenton oxidation for recalcitrant wastewater treatment", *Process Saf. Environ. Prot.,* vol. 126, pp. 119-140, 2019.
[http://dx.doi.org/10.1016/j.psep.2019.03.028]

[118]   M.B. Kasiri, N. Modirshahla, and H. Mansouri, "Decolorization of organic dye solution by ozonation; Optimization with response surface methodology", *Inter. J. Ind. Chem.,* vol. 4, no. 1, p. 3, 2013.
[http://dx.doi.org/10.1186/2228-5547-4-3]

[119]   J.P. Gould, and K.A. Groff, "The kinetics of ozonolysis of synthetic dyes", *Ozone Sci. Eng.,* vol. 9, no. 2, pp. 153-164, 1987.
[http://dx.doi.org/10.1080/01919518708552400]

[120]   J.P. Guin, Y.K. Bhardwaj, D.B. Naik, and L. Varshney, "Evaluation of efficiencies of radiolysis, photocatalysis and ozonolysis of modified simulated textile dye waste-water", *RSC Advances,* vol. 4, no. 96, pp. 53921-53926, 2014.
[http://dx.doi.org/10.1039/C4RA10304A]

[121]   A.B.C. Alvares, C. Diaper, and S.A. Parsons, "Partial oxidation by ozone to remove recalcitrance from wastewaters--a review", *Environ. Technol.,* vol. 22, no. 4, pp. 409-427, 2001.
[http://dx.doi.org/10.1080/09593332208618273] [PMID: 11329804]

[122] T.J. Fisher, and P.H. Dussault, "Alkene ozonolysis", *Tetrahedron,* vol. 73, no. 30, pp. 4233-4258, 2017.
       [http://dx.doi.org/10.1016/j.tet.2017.03.039]

[123] J.N. Putro, Y-H. Ju, F.E. Soetaredjo, S.P. Santoso, and S. Ismadji, "Biosorption of dyes", In: *Green Chemistry and Water Remediation: Research and Applications.,* S.K. Sharma, Ed., Elsevier, 2021, pp. 99-133.
       [http://dx.doi.org/10.1016/B978-0-12-817742-6.00004-9]

[124] A.M. Elgarahy, K.Z. Elwakeel, S.H. Mohammad, and G.A. Elshoubaky, "A critical review of biosorption of dyes, heavy metals and metalloids from wastewater as an efficient and green process", *Cleaner Engineering and Technology,* vol. 4, p. 100209, 2021.
       [http://dx.doi.org/10.1016/j.clet.2021.100209]

[125] K. Chojnacka, "Biosorption and bioaccumulation – the prospects for practical applications", *Environ. Int.,* vol. 36, no. 3, pp. 299-307, 2010.
       [http://dx.doi.org/10.1016/j.envint.2009.12.001] [PMID: 20051290]

[126] L. Smoczyński, B. Pierożyński, and T. Mikołajczyk, "The effect of temperature on the biosorption of dyes from aqueous solutions", *Processes,* vol. 8, no. 6, p. 636, 2020.
       [http://dx.doi.org/10.3390/pr8060636]

[127] M. Asgher, "Biosorption of reactive dyes: A review", *Water Air Soil Pollut.,* vol. 223, no. 5, pp. 2417-2435, 2012.
       [http://dx.doi.org/10.1007/s11270-011-1034-z]

[128] D. Park, Y.S. Yun, and J.M. Park, "The past, present, and future trends of biosorption", *Biotechnol. Bioprocess Eng.; BBE,* vol. 15, no. 1, pp. 86-102, 2010.
       [http://dx.doi.org/10.1007/s12257-009-0199-4]

[129] Z. Aksu, "Reactive dye bioaccumulation by Saccharomyces cerevisiae", *Process Biochem.,* vol. 38, no. 10, pp. 1437-1444, 2003.
       [http://dx.doi.org/10.1016/S0032-9592(03)00034-7]

[130] G. Dönmez, "Bioaccumulation of the reactive textile dyes by Candida tropicalis growing in molasses medium", *Enzyme Microb. Technol.,* vol. 30, no. 3, pp. 363-366, 2002.
       [http://dx.doi.org/10.1016/S0141-0229(01)00511-7]

[131] S. Sadettin, and G. Dönmez, "Bioaccumulation of reactive dyes by thermophilic cyanobacteria", *Process Biochem.,* vol. 41, no. 4, pp. 836-841, 2006.
       [http://dx.doi.org/10.1016/j.procbio.2005.10.031]

[132] I. Ihsanullah, A. Jamal, M. Ilyas, M. Zubair, G. Khan, and M.A. Atieh, "Bioremediation of dyes: Current status and prospects", *J. Water Process Eng.,* vol. 38, p. 101680, 2020.
       [http://dx.doi.org/10.1016/j.jwpe.2020.101680]

[133] S.K. Sharma, *Bioremediation: A Sustainable Approach to Preserving Earth's Water.* CRC Press, 2019.
       [http://dx.doi.org/10.1201/9780429489655]

[134] K. Vikrant, B.S. Giri, N. Raza, K. Roy, K.H. Kim, B.N. Rai, and R.S. Singh, "Recent advancements in bioremediation of dye: Current status and challenges", *Bioresour. Technol.,* vol. 253, pp. 355-367, 2018.
       [http://dx.doi.org/10.1016/j.biortech.2018.01.029] [PMID: 29352640]

[135] R. Khan, V. Patel, and Z. Khan, "Bioremediation of dyes from textile and dye manufacturing industry effluent", In: *Abatement of Environmental Pollutants.,* P. Singh, A. Kumar, A. Borthakur, Eds., Elsevier, 2020, pp. 107-125.
       [http://dx.doi.org/10.1016/B978-0-12-818095-2.00005-9]

[136] P. Somu, S. Narayanasamy, L.A. Gomez, S. Rajendran, Y.R. Lee, and D. Balakrishnan, "Immobilization of enzymes for bioremediation: A future remedial and mitigating strategy", *Environ.*

Res., vol. 212, no. Pt D, p. 113411, 2022.
[http://dx.doi.org/10.1016/j.envres.2022.113411] [PMID: 35561819]

[137] A. Kandelbauer, and G.M. Guebitz, "Bioremediation for the Decolorization of Textile Dyes — A Review", In: *Environmental Chemistry: Green Chemistry and Pollutants in Ecosystems.,* E. Lichtfouse, J. Schwarzbauer, D. Robert, Eds., Springer Berlin Heidelberg: Berlin, Heidelberg, 2005, pp. 269-288.
[http://dx.doi.org/10.1007/3-540-26531-7_26]

[138] M. Ajaz, S. Shakeel, and A. Rehman, "Microbial use for azo dye degradation—a strategy for dye bioremediation", *Int. Microbiol.,* vol. 23, no. 2, pp. 149-159, 2020.
[http://dx.doi.org/10.1007/s10123-019-00103-2] [PMID: 31741129]

[139] G. Ravindiran, G.P. Ganapathy, J. Josephraj, and A. Alagumalai, "A Critical insight into biomass derived biosorbent for bioremediation of dyes", *ChemistrySelect,* vol. 4, no. 33, pp. 9762-9775, 2019.
[http://dx.doi.org/10.1002/slct.201902127]

[140] N.S. Kv, "Removal of dyes from industrial effluents using bioremediation technique", In: *Strategies and Tools for Pollutant Mitigation: Avenues to a Cleaner Environment.,* J. Aravind, M. Kamaraj, M. Prashanthi Devi, S. Rajakumar, Eds., Springer International Publishing: Cham, 2021, pp. 173-194.
[http://dx.doi.org/10.1007/978-3-030-63575-6_9]

[141] M. Premaratne, G.K.S.H. Nishshanka, V.C. Liyanaarachchi, P.H.V. Nimarshana, and T.U. Ariyadasa, "Bioremediation of textile dye wastewater using microalgae: current trends and future perspectives", *J. Chem. Technol. Biotechnol.,* vol. 96, no. 12, pp. 3249-3258, 2021.
[http://dx.doi.org/10.1002/jctb.6845]

[142] S. Unnikrishnan, N. Bora, and K. Ramalingam, "Bioremediation: A Logical Approach for the Efficient Management of Textile Dye Effluents", In: *Handbook of Research on Resource Management for Pollution and Waste Treatment.,* A.C. Affam, E.H. Ezechi, Eds., IGI Global: Hershey, PA, USA, 2020, pp. 294-317.
[http://dx.doi.org/10.4018/978-1-7998-0369-0.ch013]

[143] J. Paul Guin, Y.K. Bhardwaj, and L. Varshney, "Mineralization and biodegradability enhancement of Methyl Orange dye by an effective advanced oxidation process", *Appl. Radiat. Isot.,* vol. 122, pp. 153-157, 2017.
[http://dx.doi.org/10.1016/j.apradiso.2017.01.018] [PMID: 28161647]

[144] L. Bardi, and M. Marzona, "Factors affecting the complete mineralization of azo dyes", In: *Biodegradation of Azo Dyes.,* H. Atacag Erkurt, Ed., Springer Berlin Heidelberg: Berlin, Heidelberg, 2010, pp. 195-210.
[http://dx.doi.org/10.1007/698_2009_50]

[145] M. Aslam, I.M.I. Ismail, S. Chandrasekaran, H.A. Qari, and A. Hameed, "How the Dyes are degraded/mineralized in a photocatalytic system? the possible role of auxochromes", *Water Air Soil Pollut.,* vol. 226, no. 3, p. 70, 2015.
[http://dx.doi.org/10.1007/s11270-015-2301-1]

CHAPTER 2

Toxicity Analysis of Dyes

Arumugam Girija[1,*] and **Paulpandian Muthu Mareeswaran**[2]

[1] *Department of Chemistry, Velumanoharan Arts & Science College for Women, Ramanathapuram-623540, Tamil Nadu, India*

[2] *Department of Chemistry, College of Engineering, Anna University, Chennai-600025, Tamil Nadu, India*

Abstract: In the textile sector, synthetic dyes are crucial. However, dyes pose a serious threat to all organisms because of their toxicity. Environmental concerns have grown over the non-selective and excessive usage of these dyestuffs. These colours have the potential to be harmful in terms of behaviour, biology, chemistry, physicality, and radiation. The toxicity of the dyes can be classified as acute (short-term effect) or chronic (long-term damage). In order to establish criteria for the regulation of dyes when they come into contact with humans and other living things, toxicity analyses of dyes are therefore required. In a toxicology study, the reaction of an organism to a specific dye at different concentrations is compared to the reaction of the same organisms not exposed to the dye. The toxic effects of an experimental substance are revealed by toxicity testing on numerous biological systems. The producers utilize this evaluation to determine the dye's toxicity and whether it has carcinogenic or non-carcinogenic effects.

Keywords: Acute toxicity, Chronic toxicity, Dose descriptors, Dose-response relationship, Synthetic dyes, Sub-acute toxicity.

INTRODUCTION

Dyes are sophisticated organic compounds that are used to colour products in a variety of industries, including the textile, printing, rubber, cosmetics, plastics, and leather industries. Dyes can be derived from synthetic or natural sources. Natural dyes are made without the use of chemicals from naturally existing substances such as plants (such as indigo), insects (such as cochineal beetles), animals (such as shellfish), and minerals (such as ferrous sulphate). Synthetic dyes, such as Congo red and methyl orange are synthesised in laboratories. Due to factors including inexpensive production, simple availability, straightforward application, greater colour stability, resistance to light, pH variations, oxidation,

---

** **Corresponding author Arumugam Girija:** Department of Chemistry, Velumanoharan Arts & Science College for Women, Ramanathapuram-623540, Tamil Nadu, India; E-mail: girijabalaji99@gmail.com*

Paulpandian Muthu Mareeswaran & Jegathalaprathaban Rajesh (Eds.)

etc., synthetic dyes have attracted a lot of attention and have practically supplanted natural dyes. Both humans and the environment are harmed by synthetic dyes. Lead, mercury, chromium, copper, toluene, benzene, and others are some of the compounds that are present in synthetic dyes [1 - 5]. The human body may suffer from serious consequences if exposed to high concentrations of certain drugs.

They will result in alterations in immunoglobulin levels, allergic dermatoses, and respiratory illnesses. Certain azo dyes may be mutagenic. Even the chemicals used to make colours have been shown to be poisonous, cancer-causing, or occasionally explosive [5 - 8]. For instance, aniline, which is used to make azo colours, is toxic and extremely combustible. Additionally, the chemicals used in the dying process are dangerous. For instance, dioxin may be carcinogenic and affect hormones. Heavy metals including chromium, copper, and lead are inherently carcinogenic. Table **1** lists some of the chemicals used in the dying industry along with their average usage per month [9, 10].

Table 1. Commonly used chemicals in the dying industry.

S. No.	Chemicals Used	Quantity Kg/month
1.	Acetic acid	1611
2.	Ammonium sulphate	858
3.	PV acetate	954
4.	Wetting agent	125
5.	Caustic Soda	6212
6.	Softener	856
7.	Organic Resin	5115
8.	Formic acid	1227
9.	Soap	154
10.	Hydro sulphites	6563
11.	Hydrogen Peroxide	1038
12.	Levelling & Dispersing Agents	547
13.	Oxalic acid	471
14.	Polyethylene emulsion	117
15.	Sulphuric acid	678

Azo dyes are used extensively in the textile, pharmaceutical, printing, food, and other sectors, making up close to 70% of all synthetic colours. Numerous businesses release toxic effluents that contain azo dyes, which have a negative impact on ecosystem health, soil fertility, aquatic life, and resource availability. They are poisonous to aquatic creatures (fish, algae, bacteria, *etc.*), much like they are to animals (lethal impact, genotoxicity, mutagenicity, and carcinogenicity). They are not easily degradable in the environment, and typical wastewater treatment methods do not eliminate them from sewage. It has long been known that benzidine-based dyes can cause tumours in a variety of laboratory animals and can cause cancer of the urinary bladder in humans [11].

TOXICITY

The capacity of a material to harm living things is referred to as its toxicity. Even if a harmful agent is given in extremely small levels, the cells will still suffer long-term damage. Only when the less harmful drug is consumed in high quantities will it have an impact on the organism. As a result, it is important to define a substance's toxicity in terms of the amount of the substance administered or absorbed, the mode of administration (such as injection, ingestion, or inhalation), the time distribution (such as a single dose or repeated doses), the type and severity of the injury, and the amount of time needed to cause the injury [12].

Toxic Effects

Acute impacts and chronic effects are two different types of toxic reactions that might take place after repeated exposures over an extended period. The level of severity may also vary, and they may only affect a few or all body parts. Acute effects occur quickly following a single encounter; examples include food poisoning, inhaling chlorine spill fumes, *etc.* Examples of immediate consequences include sweating, nausea, paralysis, and death. After repeated exposure over an extended period, chronic impacts may develop; for instance, smoking cigarettes, consuming foods with low amounts of pollutants, inhaling polluted air, *etc.* Chronic impacts include cancer, organ damage, problems getting pregnant, and nervous system impairment, for instance. Chronic impacts may have cancer-causing or non-cancerous effects [13, 14].

Toxicological Field Studies

The nature of the poisonous substance, the nature of the environmental pollution it causes, and the nature of the species present in the specific environment are all taken into consideration during the toxicological testing on species. Testing wild species kept in cages in field circumstances can potentially result in several issues.

A big site with a variety of test animals is needed for species assessment. Additional information gathered with lab animals is derived using a toxic assessment of dyes from outdoor trials. This supports the validity of extrapolating experimental findings to the ecosystem [15].

Priorities in the Selection of Toxic Chemicals for Testing

Prior to manufacture, all new chemicals must undergo a safety study; however, considering the prevalence of poisonous dyes that are more dangerous to human health, priority should be given to those substances that come into direct contact with people. Priority should be given to dangerous substances with high suspicions of acute, chronic, or delayed harm. Numerous halogenated dyes are resistant to metabolism, and metabolism by bacteria, in particular, will have a high environmental persistence. In determining priorities, physicochemical characteristics of the hazardous dyes are also crucial. A persistent, fat-soluble substance's biomagnification, for instance, will contaminate human food supplies as well as wildlife at higher levels of food chains. Through adsorption (air or water), toxic chemicals on soil particles may be transported to distant locations from the point of application. These compounds require top priority as well. Priority should be given to substances that (i) indicate or raise suspicions of posing a risk to human health as well as the kind and intensity of potential adverse effects on health. (ii) The extent of production and usage, (iii) the potential for environmental accumulation, (iv) the kind and amount of the population that will likely be exposed [16].

The Extent of the Toxicity Requirement

Starting with its synthesis, the initial examination of toxicity assessment should be conducted. The assessment should also take into account general exposure, occupational risks, and air, water, and food pollution. The information on the harmful effects of raw materials, as well as other compounds utilized or created as intermediates during the process, is provided by the evaluation data of toxic chemicals. To decrease the harmful effect, alternative technological methods or products may be selected with the use of toxicological evaluation [16].

An evaluation of the toxicity of a substance is used to determine the hazardous effect. One of the key elements of risk assessment is toxicity assessment [13]. Mere The sheer existence of a material does not necessarily imply that it is dangerous. But after being limited in value, they could become toxic. For instance, when administered at the right dosage, some medicinal drugs that are acutely poisonous can be helpful. For instance, excessive amounts of oxygen, vitamin D, table salt, and water are poisonous. As a result, toxicity evaluation considers the kind and severity of harm brought on by various dosages of a chemical [16].

Toxicity Assessment

The dose-response principle is the foundation for toxicity evaluation. This has distinct applications for determining acute and chronic toxicity. To find out what kind of effects a high dose can have, experiments are carried out. The more poisonous the colour is, the smaller the dose required to produce the effect. Except for carcinogens, it is thought that every substance has a toxicological threshold value below which it is non-toxic [17].

There is no upper limit for carcinogenic substances because even the tiniest dose might have deadly effects. Although dose-response studies are the most popular technique for assessing toxicity, acute and chronic effects are studied using different approaches. To assess the acute toxic effect, researchers use LD_{50} experiments or monitoring of unintentional exposures. Animal dose-response studies are used to assess the chronic toxicity of substances. The carcinogenesis bioassay is used to assess the effects of carcinogens [18].

DOSE

The amount of chemical supplied, represented as a dose per unit of body weight, is called a dose. The dose may be given orally, topically, or through the respiratory system, but the adsorbed dose may differ from the delivered dose. The dose supplied, the concentration in cells, and the concentration in faeces or exhaled air can all be used to determine how much the dose is in body cells. It is necessary to postulate models that describe the absorption, distribution, retention, biotransformation, and excretion of the hazardous dyes as a function of time in order to employ the three forms of information mentioned above. If the site of harmful action is known, the cell dose calculation may be accurate [15]. Toxic chemical levels in the blood are a sign of chemical absorption. Only when there is a clear relationship between the blood chemical concentration and the concentration at the sites of action can it be used as an indicator.

Toxicological Dose Descriptors

Dose descriptors are used to show the connection between a particular chemical effect and the dose at which it occurs. In toxicological investigations, these dosage descriptors are determined and typically expressed as LC_{50}, LD_{50}, NOAEL, NOAEC, T_{25}, BMD, EC_{50}, NOEC, DT_{50}, *etc* [16, 17].

LD_{50} *(Lethal Dose 50%)*

The term LD_{50} value" refers to the dose at which 50% of testing animals or species perish after receiving a single treatment, service, or exposure. This is quite helpful in figuring out acute poisoning [18].

LC_{50} *(Lethal Concentration 50%)*

The concentration at which 50% of testing animals or species perish after one treatment or exposure during a specific time period is known as the LC_{50}. This is quite helpful in figuring out acute toxicity. This is highly useful in figuring out the LC_{lo} value, which is the lowest dose at which the testing animal or species can die [18].

No Observed Adverse Effect Level (NOAEL)

It is the level of exposure that, between the exposed population and its suitable control, does not significantly increase the frequency or severity of adverse effects. NOAEL is typically calculated from studies on reproductive toxicity and repeated dosage toxicity. NOAELs, such as the derived no-effect level (DNEL), occupational exposure limit (OEL), and acceptable daily intake (ADI), are used to determine the threshold safety exposure dose to humans. NOAEL is measured in mg/Kg bw/day or ppm. Lower systemic toxicity or lower chronic toxicity is indicated by higher NOAEL or NOAEC [19].

Lowest Observed Adverse Effect Level (LOAEL)

The lowest exposure level at which adverse effects are significantly more frequent or severe in the exposed population compared to the relevant control group is known as the Lowest Observed Adverse Effect Level (LOAEL). As with the derived no-effect level (DNEL), LOAEL are frequently used to determine threshold safety exposure doses for humans [20].

T_{25} *and* BMD_{10}

Certain substances have the potential to be carcinogenic even at low exposure levels. The typical NOAEL cannot be found for risk assessment in several carcinogenicity investigations. T_{25} and BMD_{10} are employed as a result.

$\underline{T_{25}}$

The sustained dosage rate that, after adjusting for spontaneous occurrence, results in 25% of animal tumours in a given tissue throughout the course of life of a species [21].

BMD_{10}

BMD_{10} is a benchmark dose that is thought to contribute 10% of the tumours of the animals at a certain tissue after adjusting for spontaneous occurrence over the course of the lifetime of the species [21].

Median Effective Concentration (EC_{50})

The EC_{50} in ecotoxicity is the substance concentration that results in a 50% reduction in the growth of test organisms of either algae (EbC50), the rate of algae growth (ErC_{50}), or *Daphina* immobilization. mg/L is the unit of measurement for EC_{50} [22].

No Observed Effect Concentration (NOEC)

It is the concentration in an environmental compartment, such as water, air, soil, *etc.*, below which it is unlikely to have an unfavourable effect. Typically, investigations on chronic aquatic toxicity and terrestrial toxicity are used to acquire it. NOEC is measured in mg/L as its unit [23].

DT_{50}

Half-life (DT_{50}) is the amount of time it takes for a material to degrade to half its original amount in an environmental compartment (such as water, soil, or air). It is employed to gauge a substance's persistence. "Per day" is the unit DT_{50}. Table **2** lists general classification criteria for toxicity and the connection between LC, LD, and EC and toxicity rating [24].

Table 2. The toxicity parameters and their activity related to percentage.

LD_{50}mg/Kg	LC_{50}mg/L	EC_{50}	Toxicity Rating
>5000	>100	>100%	Relatively not acutely toxic
500-5000	10-100	10-100%	Minor acutely toxic
50-500	1-10	1-10%	Moderately acutely toxic
<50	<1	<1%	Very acutely toxic

Effect and Response

The terms "effect" and "response" are used to denote biological changes in a person or a population that are related to exposure or dose; thus, "effect" refers to the biological change and "response" refers to the percentage of the population that exhibits an outlined effect; in other words, "response" means the incidence rate of an effect. Although the poisonous material has an impact on the entire body, its effects are also focused in a specific organ, which may lead to that organ's dysfunction or influence it through a specific disease. The zone of specific action (Z_{sp}), which is defined as the ratio between the threshold dose of an acute effect at the level of the entire organism and the threshold dose of an acute effect at the specific organ, can be used to indicate the specificity of acute action. The toxic activity is non-specific if Z_{sp} is less than one, and specific if Z_{sp} is more than one [25].

Dose Effect and Dose-Response Curves

The association between the dose and the size of the graded effect within the test species is provided by dose-effect curves (Fig. **1**). They could have linear or non-linear curves. The relationship between dose and the percentage of test species exhibiting a quantal response is shown by dose-response curves. Dose-response curves are often S-shaped. However, due to variations in the experimental settings, such as how the dye is dispersed over time in the environment, the shape of the dose-response curve may alter for the same dye and the same animal species. The amount of dosage is evaluated as a function of concentration and duration when assessing human exposure to hazardous dyes. In some circumstances, the concentration may be constant, in which case the correlations between the time effects and time responses will be identical to those between the dosage effects and dose responses. However, in the majority of cases, the dose-time-concentration connection must be taken into account because the dye concentration may change with exposure time to a specific concentration [26].

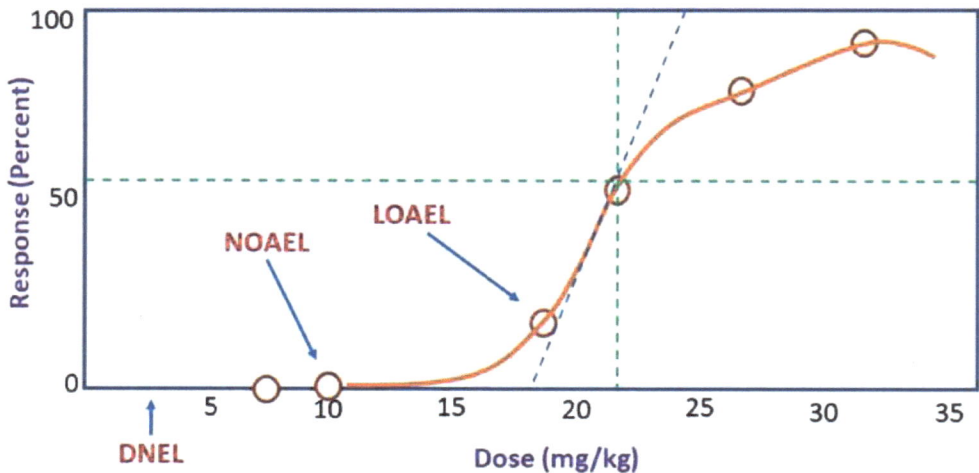

Fig. (1). Dose response curves.

According to Haber's Rule, the Law of Toxicology states that the extent of a toxic effect depends on the total exposure, *i.e.*, exposure concentration (c) rate times the duration time (t) of exposure (c × t).

$$Ct^m = K \tag{1}$$

Where C- Concentration; t -Time; K- Constant

This rule, within constraints, is commonly employed in setting exposure guidelines for toxic substances. The exponent m is a kind of variable. For inhaled toxic dyes, m generally includes a value between 0 and 1, whereas for carcinogenic dyes m is typically between 1 and 5 [27].

Toxic Effects Due to Combination of Dyes

When an organism is exposed to two or more harmful dyes, their combined effects may be:

(i) Independent - When the dyes have different effects or modes of action.

(ii) Additive - When the combined action of two or more dyes produces a response that is equivalent to the sum of their separate effects.

(iii) Potentiation or synergism, which goes beyond additive effects.

(iv) Antagonism or inhibition, which are less than additive.

It is possible that joint action at a deadly dose differs from that at a low level. Typically, the combined effect will be independent or additive [28].

Acute, Sub-acute, Sub-chronic and Chronic Toxic Assessment

The assessment indicated above aims to identify the impact of poisonous dyes on biological systems and to gather information on dye dose characteristics. The degree of risk to man and consequently the environment may be revealed by these statistics. In experimental toxicology, the choice of appropriate test protocols, meticulous adherence to approved experimental practices, and creative observation are of the utmost importance [29 - 31].

Acute toxicity assessment

Adverse effects that manifest quickly after the administration of a single dose or multiple doses given within 24 hours are referred to as acute toxicity. Acute toxicity studies are used to analyse the relative toxicity of the molecule, to uncover its mode of action and its specific toxic effect, and to establish the existence of species differences when data regarding the toxicity of the test agent are absent. Acute toxicity is assessed using experiments on animals known as LD_{50} studies. Milligrams of material per kilogram of body weight, or mg/kg, is how the LD_{50} is expressed. The more hazardous the material is, the lower the LD_{50} value [18 - 21]. A small number of test animals are given calculated amounts of dye to evaluate acute toxicity, and they are then monitored for 24 hours. If there are no deaths or other signs after 24 hours, the observation period is extended, and the proportion of test animal deaths is calculated. We convert this proportion to probit (probability units). The LD_{50} can be estimated from a graph that plots probit against $\log_{10}$ dosage. The following formula may also be utilized to compute it [32].

$$LD_{50} = [M_0 + M_1]/2 \qquad\qquad (2)$$

M_0 = The highest dose of test substance without mortality

M_1 = The lowest dose of test substance that gave mortality

Experimental Design

Selection of Species

The three most frequently utilized test species are the mouse, rat, and dog. The LD_{50} calculations must be made in both male and female animals, and they are presumably partly influenced by variations in liver metabolism. The age of test animals may have a significant impact on acute toxicity. Animals of varying ages are used to calculate LD_{50} values. The impact of LD_{50} may be due to differences in drug-metabolizing enzyme levels, the absence of sex hormone impacts, or decreased central nervous system sensitivity [32].

Selection of Doses

The dosages are chosen in order to determine the LD_{50} and learn more about the slope of the dose-response curve. The initial dose that will be given is selected so that no effect is shown in the target animals. Alternatively, the dose may be raised over time until it reaches a level where all the target animals die. Using the data gathered, a dose-response curve is produced from which an LD_{50} value could also be determined [32].

Method of Administration

The substances should typically be supplied through the same pathway that a guy would be exposed. Additionally, the appropriate amount of liquid or carrier should be administered, and the carrier itself should not be hazardous to the target animals. Acute dermal, ocular, and inhalation investigations may be necessary to evaluate the risk to individuals handling the poisonous material in the laboratory, even though oral exposure to humans is the most common route of exposure. Only 1% (44) of the 61 commercial dyes tested for acute toxicity on rats had a lethal dose (LD_{50}) of less than 250 mg/kg, while 92% of the products tested had low acute toxicity surpassing 2000 mg/kg. It was discovered that the fundamental dyes exhibit noticeably increased toxicity [33].

Sub-acute Toxicity

The acute toxicity test could be followed by the sub-acute toxicity test. It entails a prolonged duration of observation of test animals used for acute toxicity. Even after receiving poisonous colours for 14 days while participating in acute toxicity trials, animals did not develop mortalities. There are indications of toxicity and mortality. When a fatality is noted within the observation period of 14 days, the dose that caused the death is administered to two animals, who are then watched for 14 days. Its sub-acute toxicity at such a level may be confirmed by the observation of one animal death during the observation period. Equation (2) contains the formula for determining the subacute lethal dose.

Sub-chronic Toxicity

Data from both the acute and subacute toxicity test levels are required for sub-chronic toxicity assessment. For this test, a minimum of three dose levels are advised. The formula below is used to determine the test doses:

$$Dose = M_0 /[1 + K] \tag{3}$$

M_0 = The highest dose of test substance without mortality after the subacute test.

K = Positive real numbers starting from 0 with an increment number of 1(*i.e.*, $K = 0$, $K = 1$ and $K = 2$ for three respective dose levels).

Animals are given the test doses, and daily observations are made of each animal for the duration of the study. Both toxicity and mortality signs are noted. The formula to calculate the Lethal dose using these data is:

$$LD_{50(subchronic)} = [M_0 + M_1]/2 \tag{4}$$

M_0 = The highest dose of test substance without mortality after the subchronic test. M_1 = The lowest dose of test substance that gave mortality after the subchronic test.

Chronic Toxicity Assessment

Chronic toxicity may be divided into two categories. (i) Carcinogenic toxicity, and (ii) Non-carcinogenic toxicity.

Assessment of Non-carcinogenic Toxicity

By giving the chemical to the testing animals in varied amounts and seeing the results, non-carcinogenic toxicity is frequently evaluated. This aids in determining the lowest dose necessary to produce observable effects. Researchers provide the dye in a variety of modest dosages over the course of a lifetime to participants in the dose-response experiments. The animals are periodically inspected, and finally, an autopsy is performed to look for behavioural changes, changes in the concentration of vital bodily substances, and damage to any organs. LOEL is expressed in parts per million (ppm) or milligrams of substance per kilogram of body weight [20].

Selection of Species and Duration of Studies

Animals are exposed to dyes for a prolonged amount of time throughout their test animal lives in order to determine their chronic toxicity. Although rodents are frequently employed as test species, other large animals, such as dogs and monkeys, are also used. Utilizing a sufficient number of test animals is also crucial to achieving a statistically sound design. As the target animals must be exposed throughout a lengthy lifespan, it is essential to begin the exposure process early in life. The greatest dose at which no effects are seen (NOEL) is considered during experimentation as a safe level for the hazardous substances present in the species. But NOEL is not necessarily the safe level for humans, because humans could also be more or less sensitive to the toxic substance than the animals studied [19].

Human Sensitivity and Variability

Humans are assumed to be more sensitive than animals for the sake of calculating the safe level, however, this is not always the case. This may guarantee that different species will absorb, process, and excrete the drug at different rates. Humans tend to react more like monkeys do, but occasionally they may act differently. Dogs and humans will both react the same way to nitrobenzene, while monkeys will not respond at all. As a result, reactions to a certain harmful chemical may differ between species [34].

Assessment of Carcinogenic Toxicity

The evaluation of carcinogenic toxicity is distinct from the evaluation of non-carcinogenicity. Millions of test animals must be used for scientists to accurately examine carcinogenic toxicity. It takes a significant amount of financial and laboratory resources to assess carcinogenicity. Additionally, because most dyes have little toxicity, it takes a lot of experience to interpret and apply the dose-response relationship when testing at the maximum tolerated level. Using high-dose research on lab animals, the carcinogenic effects of dyes will be investigated in this way to screen for even the uncommon case of cancer [35].

Methodology

Scientists can determine the carcinogenic hazard of substances by subjecting test animals to dyes in an effort to detect even the most sporadic cases of cancer. Over the course of a lifetime, test animals are exposed to various high dosages of dyes regularly. The development of cancer is then inspected in the animals. If cancer is discovered, the scientists estimate the rate of growth of cancer cells even at lower doses using the data and mathematical models that are now accessible. The model also calculates the impact of size and sensitivity variations between a group of test animals and humans [35].

A set of equations that replicate real-world conditions and forecast future events could be referred to as a mathematical model. It is challenging to determine what would happen to humans exposed to low amounts prevalent in the environment when considering toxicity. Therefore, mathematical models are created in order to apply the information from animal research to the human condition. The choice of models has a significant impact on the toxicity assessment since different findings will be obtained when using various models on the same set of data. For instance, benzene and similar dyes were subjected to a rat and mouse carcinogenesis bioassay. High doses of *leukemia* afflicted both species. A benzene dose of 1 mg/kg/day resulted in three malignancies per 100 people exposed daily for their entire lives for that level, according to extrapolating cancer incidence from high dose to low dose from rodents to humans [35].

MATHEMATICAL MODELS

A mathematical model is a collection of equations that simulates a real-world scenario and forecasts future events. It is challenging to predict what would happen to humans exposed to low concentrations present in the environment when considering toxicity. Therefore, mathematical models are created to translate the results of animal studies to the human condition. The choice of models has a significant impact on the assessment of toxicity since different results will be obtained when several models are applied to the same data [36].

Types of Models

Two different mathematical models are generally used for assessing carcinogenic toxicity. They are: (i) Threshold Model; and (ii) Non-Threshold model [36].

Threshold Model

The threshold model is predicated on the idea that numerous harmful material exposures are required before a threshold of exposure is achieved [37].

Non-Threshold Model

The Non-Threshold Model is predicated on the idea that even one carcinogen molecule can result in the disease. One-Hit Model is another name for this model. Scientists typically use non-threshold models to evaluate the toxicity of carcinogens. These provide a chemical with an estimated cancer potency that is higher than what threshold models would [38].

TOXICITY EVALUATION OF EFFLUENTS FROM THE DYE INDUSTRY USING DAPHNIA MAGNA

Recent years have seen a rise in the usage of the crustacean *Daphnia magna* (Fig. 2) as a sensor organism and the LC_{50} criterion as a standard for assessing the toxicity of textile effluent [39]. Due to their rapid reproduction, sensitivity to chemical environments, and important ecological role in the aquatic food chain by acting as an intermediary between primary producers and fish, *Daphnia magna* have been recommended as a good organism to test the effluent toxicity in industrial wastewater and textile wastewater containing dyes [40 - 42]. There is also proof of a close connection between rats and *Daphnia magna*, which is now used in toxicity testing in place of rats [43].

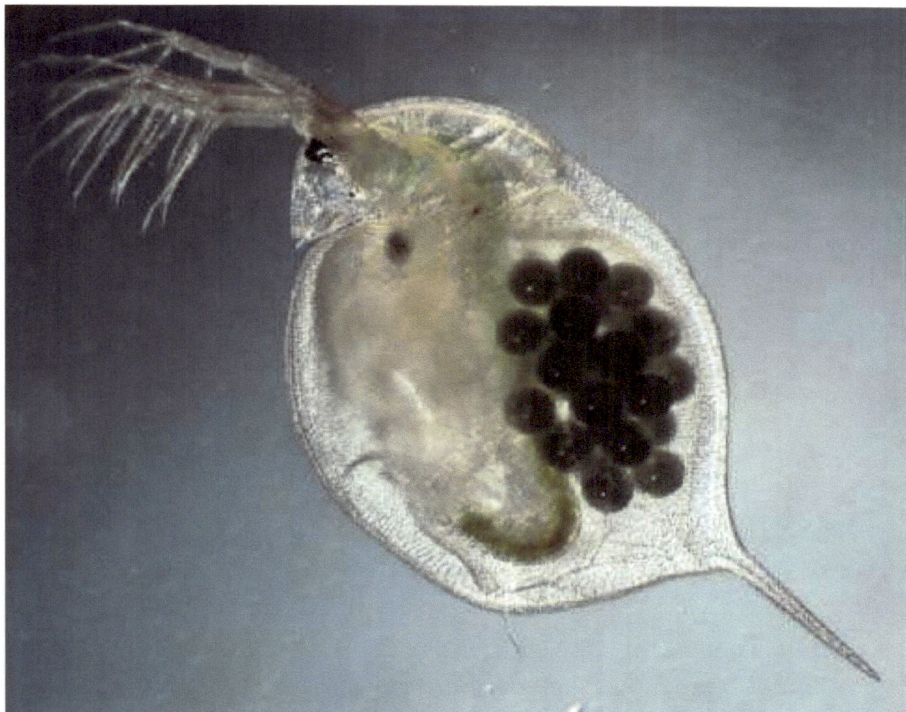

Fig. (2). *Daphnia Magna* used for toxicity analysis.

CONCLUSION

The evaluation of toxicity provides both quantitative and qualitative data on carcinogens. If the toxicity evaluation is based on animal studies, it is necessary to extrapolate the degree of harm to people by utilizing a mathematical model and numerous suppositions. As a result, the toxicity evaluation only provides an estimation of the detrimental impact on human health. The actual negative impact on the biological system can be predicted using this research because the toxicity evaluation depends on dose-response studies on many animal species, epidemiological studies, and *in vitro* studies. The majority of synthetic dyes are known carcinogens, and their presence will have a negative impact due to chemical interactions with biomolecules, interactions with light, and the subsequent formation of singlet oxygen. Because of this, the toxicity analysis of dyes is crucial for dye compounds.

REFERENCES

[1]   E. Lanciotti, S. Galli, A. Limberti, and L. Giovannelli, "Ecotoxicological evaluation of wastewater treatment plant effluent discharges: A case study in Prato (Tuscany, Italy)", *Ann. Ig.,* vol. 16, no. 4, pp. 549-558, 2004.

[PMID: 15366513]

[2]   E. Ellouze, N. Tahri, and R.B. Amar, "Enhancement of textile wastewater treatment process using Nanofiltration", *Desalination,* vol. 286, pp. 16-23, 2012.
[http://dx.doi.org/10.1016/j.desal.2011.09.025]

[3]   G. Eremektar, H. Selcuk, and S. Meric, "Investigation of the relation between COD fractions and the toxicity in a textile finishing industry wastewater: Effect of preozonation", *Desalination,* vol. 211, no. 1-3, pp. 314-320, 2007.
[http://dx.doi.org/10.1016/j.desal.2006.02.096]

[4]   K.P. Sharma, S. Sharma, S. Sharma, P.K. Singh, S. Kumar, R. Grover, and P.K. Sharma, "A comparative study on characterization of textile wastewaters (untreated and treated) toxicity by chemical and biological tests", *Chemosphere,* vol. 69, no. 1, pp. 48-54, 2007.
[http://dx.doi.org/10.1016/j.chemosphere.2007.04.086] [PMID: 17583772]

[5]   Y. Verma, "Acute toxicity assessment of textile dyes and textile and dye industrial effluents using Daphnia magna bioassay", *Toxicol. Ind. Health,* vol. 24, no. 7, pp. 491-500, 2008.
[http://dx.doi.org/10.1177/0748233708095769] [PMID: 19028775]

[6]   V. Tigini, P. Giansanti, A. Mangiavillano, A. Pannocchia, and G.C. Varese, "Evaluation of toxicity, genotoxicity and environmental risk of simulated textile and tannery wastewaters with a battery of biotests", *Ecotoxicol. Environ. Saf.,* vol. 74, no. 4, pp. 866-873, 2011.
[http://dx.doi.org/10.1016/j.ecoenv.2010.12.001] [PMID: 21176963]

[7]   M. Chhabra, S. Mishra, and T.R. Sreekrishnan, "Combination of chemical and enzymatic treatment for efficient decolorization/degradation of textile effluent: High operational stability of the continuous process", *Biochem. Eng. J.,* vol. 93, pp. 17-24, 2015.
[http://dx.doi.org/10.1016/j.bej.2014.09.007]

[8]   Y. Mountassir, A. Benyaich, M. Rezrazi, P. Berçot, and L. Gebrati, "Wastewater effluent characteristics from Moroccan textile industry", *Water Sci. Technol.,* vol. 67, no. 12, pp. 2791-2799, 2013.
[http://dx.doi.org/10.2166/wst.2013.205] [PMID: 23787319]

[9]   S.H. Hashemi, and M. Kaykhaii, "Azo dyes: Sources, occurrence, toxicity, sampling, analysis, and their removal methods", In: *Emerging Freshwater Pollutants.,* T. Dalu, N.T. Tavengwa, Eds., Elsevier, 2022, pp. 267-287.
[http://dx.doi.org/10.1016/B978-0-12-822850-0.00013-2]

[10]   R. Kant, "Textile dyeing industry an environmental Hazard", *Nat. Sci.,* vol. 4, pp. 20-26, 2012.

[11]   R. Majumdar, W.A. Shaikh, S. Chakraborty, and S. Chowdhury, "A review on microbial potential of toxic azo dyes bioremediation in aquatic system", In: *Microbial Biodegradation and Bioremediation.,* S. Das, H.R. Dash, Eds., Elsevier, 2022, pp. 241-261.
[http://dx.doi.org/10.1016/B978-0-323-85455-9.00018-7]

[12]   D.J. Ecobichon, *The basis of toxicity testing.* CRC Press, 1997.

[13]   R.M. Joy, "Neurotoxicology: Central and Peripheral", In: *Encyclopedia of toxicology.* Academic Press, 1998.

[14]   B.C. Ventura-Camargo, and M.A. Marin-Morales, "Azo dyes: Characterization and toxicity– A review", *Text. Light Ind. Sci. Technol.,* vol. 2, pp. 85-103, 2013.

[15]   W.K. Rumbeiha, and D.B. Snider, "Veterinary Toxicology", In: *Encyclopaedia of Toxicology.,* P. Wexler, Ed., Elsevier, 2014, pp. 915-928.
[http://dx.doi.org/10.1016/B978-0-12-386454-3.00444-9]

[16]   G.A. Umbuzeiro, A.F. Albuquerque, F.I. Vacchi, M. Szymczyk, X. Sui, R. Aalizadeh, P.C. von der Ohe, N.S. Thomaidis, N.R. Vinueza, and H.S. Freeman, "Towards a reliable prediction of the aquatic toxicity of dyes", *Environ. Sci. Eur.,* vol. 31, no. 1, p. 76, 2019.
[http://dx.doi.org/10.1186/s12302-019-0258-1]

[17] D.J. Hoffman, B.A. Rattner, G.A. Burton, and J.C. Cairns, *Handbook of ecotoxicology*. Boca Raton Lewis Publishers, 1994.

[18] P.R. Parish, "Acute toxicity test", In: *In: Fundamentals of aquatic toxicology, Hemisphere* New York, 1989, p. 31.

[19] M.A. Dorato, and J.A. Engelhardt, "The no-observed-adverse-effect-level in drug safety evaluations: Use, issues, and definition(s)", *Regul. Toxicol. Pharmacol.*, vol. 42, no. 3, pp. 265-274, 2005.
[http://dx.doi.org/10.1016/j.yrtph.2005.05.004] [PMID: 15979222]

[20] R.G. Domenech, J. V. de. Julian-Ortiz, and E. Besalu, "True prediction of lowest observed adverse effect level", *Mol. Divers.*, vol. 10, pp. 159-168, 2006.
[http://dx.doi.org/10.1007/s11030-005-9007-z] [PMID: 16721628]

[21] E. Dybing, J. O'Brien, A.G. Renwick, and T. Sanner, "Risk assessment of dietary exposures to compounds that are genotoxic and carcinogenic—An overview", *Toxicol. Lett.*, vol. 180, no. 2, pp. 110-117, 2008.
[http://dx.doi.org/10.1016/j.toxlet.2008.05.007] [PMID: 18584977]

[22] S. Vaz Jr, "Toxicology in agriculture", In: *Analysis of chemical residues in agriculture.*, S. Vaz Jr, Ed., Elsevier, 2021, pp. 39-84.
[http://dx.doi.org/10.1016/B978-0-323-85208-1.00001-2]

[23] H. Walter, F. Consolaro, P. Gramatica, M. Scholze, and R. Altenburger, "Mixture toxicity of priority pollutants at no observed effect concentrations (NOECs)", *Ecotoxicology*, vol. 11, no. 5, pp. 299-310, 2002.
[http://dx.doi.org/10.1023/A:1020592802989] [PMID: 12463676]

[24] S. Beulke, and C. Brown, "Evaluation of methods to derive pesticide degradation parameters for regulatory modelling", *Biol. Fertil. Soils*, vol. 33, no. 6, pp. 558-564, 2001.
[http://dx.doi.org/10.1007/s003740100364]

[25] D. Baker, *Civilian exposure to toxic agents: Emergency medical response*. Cambridge University Press, 2012.

[26] C. Ritz, N. Cedergreen, J.E. Jensen, and J.C. Streibig, "Relative potency in nonsimilar dose–response curves", *Weed Sci.*, vol. 54, no. 3, pp. 407-412, 2006.
[http://dx.doi.org/10.1614/WS-05-185R.1]

[27] F.J. Miller, P.M. Schlosser, and D.B. Janszen, "Haber's rule: a special case in a family of curves relating concentration and duration of exposure to a fixed level of response for a given endpoint", *Toxicology*, vol. 149, no. 1, pp. 21-34, 2000.
[http://dx.doi.org/10.1016/S0300-483X(00)00229-8] [PMID: 10963858]

[28] S.M. Palácio, F.R. Espinoza-Quiñones, A.N. Módenes, C.C. Oliveira, F.H. Borba, and F.G. Silva Jr, "Toxicity assessment from electro-coagulation treated-textile dye wastewaters by bioassays", *J. Hazard. Mater.*, vol. 172, no. 1, pp. 330-337, 2009.
[http://dx.doi.org/10.1016/j.jhazmat.2009.07.015] [PMID: 19640647]

[29] Y. Gao, D. Wang, M.L. Xu, S.S. Shi, and J.F. Xiong, "Toxicological characteristics of edible insects in China: A historical review", *Food Chem. Toxicol.*, vol. 119, pp. 237-251, 2018.
[http://dx.doi.org/10.1016/j.fct.2018.04.016] [PMID: 29649491]

[30] C. Ravichandran, P.C. Badgujar, P. Gundev, and A. Upadhyay, "Review of toxicological assessment of d-limonene, a food and cosmetics additive", *Food Chem. Toxicol.*, vol. 120, pp. 668-680, 2018.
[http://dx.doi.org/10.1016/j.fct.2018.07.052] [PMID: 30075315]

[31] L. Liu, W. Wu, J. Zhang, P. Lv, L. Xu, and Y. Yan, "Progress of research on the toxicology of antibiotic pollution in aquatic organisms", *Acta Ecol. Sin.*, vol. 38, no. 1, pp. 36-41, 2018.
[http://dx.doi.org/10.1016/j.chnaes.2018.01.006]

[32] S.A. Saganuwan, "The new algorithm for calculation of median lethal dose (LD50) and effective dose

fifty (ED50) of Micrarus fulvius venom and anti-venom in mice", *Int. J. Vet. Sci. Med.,* vol. 4, no. 1, pp. 1-4, 2016.
[http://dx.doi.org/10.1016/j.ijvsm.2016.09.001] [PMID: 30255031]

[33]   S. Mani, and P. Chowdhary, *Dyes-Industrial applications and toxicity profile in Contaminants and Clean Technologies.* CRC Press, 2020.

[34]   S. Schubach, "A measure of human sensitivity in acute inhalation toxicity", *J. Loss Prev. Process Ind.,* vol. 10, no. 5-6, pp. 309-315, 1997.
[http://dx.doi.org/10.1016/S0950-4230(97)00016-8]

[35]   G.J. Nohynek, R. Fautz, F. Benech-Kieffer, and H. Toutain, "Toxicity and human health risk of hair dyes", *Food Chem. Toxicol.,* vol. 42, no. 4, pp. 517-543, 2004.
[http://dx.doi.org/10.1016/j.fct.2003.11.003] [PMID: 15019177]

[36]   F. Nani, and H.I. Freedman, "A mathematical model of cancer treatment by immunotherapy", *Math. Biosci.,* vol. 163, no. 2, pp. 159-199, 2000.
[http://dx.doi.org/10.1016/S0025-5564(99)00058-9] [PMID: 10701303]

[37]   A.Y. Yakovlev, "Threshold models of tumor recurrence", *Math. Comput. Model.,* vol. 23, no. 6, pp. 153-164, 1996.
[http://dx.doi.org/10.1016/0895-7177(96)00024-6]

[38]   W. Weber, and P. Zanzonico, "The controversial linear no-threshold model", *J. Nucl. Med.,* vol. 58, no. 1, pp. 7-8, 2017.
[http://dx.doi.org/10.2967/jnumed.116.182667] [PMID: 27754908]

[39]   Y. Verma, "Acute toxicity assessment of textile dyes and textile and dye industrial effluents using Daphnia magna bioassay", *Toxicol. Ind. Health,* vol. 24, no. 7, pp. 491-500, 2008.
[http://dx.doi.org/10.1177/0748233708095769] [PMID: 19028775]

[40]   J.S. Bae, and H.S. Freeman, "Aquatic toxicity evaluation of copper-complexed direct dyes to the Daphnia magna", *Dyes Pigments,* vol. 73, no. 1, pp. 126-132, 2007.
[http://dx.doi.org/10.1016/j.dyepig.2005.10.019]

[41]   J. Yoo, B. Ahn, J.J. Oh, T. Han, W.K. Kim, S. Kim, and J. Jung, "Identification of toxicity variations in a stream affected by industrial effluents using Daphnia magna and Ulva pertusa", *J. Hazard. Mater.,* vol. 260, pp. 1042-1049, 2013.
[http://dx.doi.org/10.1016/j.jhazmat.2013.07.006] [PMID: 23892313]

[42]   A.P.S. Immich, A.A. Ulson de Souza, S.M.A.G. Ulson de Souza, and U. Souza, "Removal of Remazol Blue RR dye from aqueous solutions with Neem leaves and evaluation of their acute toxicity with Daphnia magna", *J. Hazard. Mater.,* vol. 164, no. 2-3, pp. 1580-1585, 2009.
[http://dx.doi.org/10.1016/j.jhazmat.2008.09.019] [PMID: 18976858]

[43]   K. Enslein, T.M. Tuzzeo, H.H. Borgstedt, B.W. Blake, and J.B. Hart, "Prediction of Rat Oral LD50 From Daphnia Magna LC50 and Chemical Structure", In: *QSAR Environmental Toxicology II.,* K.L.E. Kaiser, Ed., , 1987.
[http://dx.doi.org/10.1007/978-94-009-3937-0_9]

CHAPTER 3

Utility of Nanotechnology in Dye Degradation

Seemesh Bhaskar[1,2] and **Sai Sathish Ramamurthy**[1,*]

[1] *STAR Laboratory, Central Research Instruments Facility (CRIF), Department of Chemistry, Sri Sathya Sai Institute of Higher Learning, Prasanthi Nilayam, Puttaparthi, Anantapur, Andhra Pradesh, India*

[2] *Department of Chemistry, Indian Institute of Technology (IIT) Bombay, Powai, Mumbai-400076, Maharashtra, India*

Abstract: Plasmonic nanoparticles and low-dimensional graphene-based derivatives are increasingly used for decolourization and degradation of harmful organic pollutants. However, the utility of their hybrid compositions synthesized *via* low-cost routes is rarely discussed. Our research examines the efficiency of surfactant-free nanomaterials and their composites with graphene oxide towards the degradation of four important textile and laser dyes, namely: Rhodamine B (RB), Methylene blue (MB), Sulforhodamine 101 hydrate (SR) and Fluorescein (FS). The surfactant-free metal-graphene oxide nanocomposites are engineered in two different techniques: (i) laser ablation mediated synthesis (LAMS) and (ii) multifunctional soret nano-assemblies (MSNAs). On account of the hybridized plasmonic effects from the large charge density oscillations in plasmonic nanoparticles and π-plasmons of graphene oxide, intriguing results are obtained and discussed in this chapter. The synergistic interplay and electron relay between the π-plasmons of graphene oxide and that of organic dyes (π-π stacking), in the vicinity of the plasmonic nanocomposites, significantly enhances the performance of the engineered nanomaterials toward dye degradation. The dye-degradation of xenobiotic pollutants demonstrated here opens a new door for the development of a broad spectrum of low-cost surfactant-free nanocomposites for environmental remediation. This study presents a futuristic insight to explore the synergy of low-dimensional and plasmonic nanomaterials constituting elements from different parts of the periodic table to accomplish dye degradation and related applications.

Keywords: Dye degradation, Environmental remediation, Electron relay, Laser ablation, Low-cost surfactant-free, Plasmonic nanocomposites, Soret nano-assemblies, π-plasmons.

* **Corresponding author Sai Sathish Ramamurthy:** STAR Laboratory, Central Research Instruments Facility (CRIF), Department of Chemistry, Sri Sathya Sai Institute of Higher Learning, Prasanthi Nilayam, Puttaparthi, Anantapur, Andhra Pradesh, India; E-mail: rsaisathish@sssihl.edu.in

Paulpandian Muthu Mareeswaran & Jegathalaprathaban Rajesh (Eds.)
All rights reserved-© 2023 Bentham Science Publishers

INTRODUCTION

The rapid increase in the growth and expansion of industries globally has not only benefitted mankind in terms of technological revolution but has also impacted the environment deleteriously [1 - 5]. Air, water and soil pollution is a worldwide issue that endangers the health and livelihood of terrestrial as well as aquatic life systems. Unfortunately, the release of non-treated and/or incompletely treated waste has resulted in hazardous effects on living organisms. Unjustifiable exploitation and overwhelming toxification of water bodies across the world are currently the 'hot issues' regarding water management [3 - 6]. In order to address the ever-increasing demand for pure water sources (due to population growth), methodologies, such as prevention and revalorization (recycle and reuse), need to be employed [1]. Industrial sectors relevant to dye applications are known among the highly polluting sectors in terms of the composition of the effluents and the magnitude of contamination. Generally derived from coal tar and petroleum intermediates, commercially obtainable synthetic dyes are more than 100 000 in number with a yearly production of more than 7×10^5 tons making India the second largest exporter of dyes after China [1 - 5]. While the World Bank attributes 17–20% of the water pollution to the textile finishing and dyeing industries, earlier reports underline that around 15% of the non-decomposable textile dyes are discharged into natural water bodies [4 - 6].

By and large, textile, food, leather, paper, pharmaceutical, cosmetics, paint, and printing industries massively use organic dyes and pigments in their manufacturing process. Basically, these industries use large amounts of water in fixing, dyeing, and washing procedures [6 - 10]. The emerging issue in this regard is the continuous disposal of dyes along with industrial wastes into the open environment. On account of their high solubility in water and difficulty in removal of such dyes *via* simple routes, they enter the surface and groundwater bodies, perpetually causing serious risks. Besides, from the chemistry point of view, these dyes are impervious to degradation by heat, light, or natural cleansing agents in environmental water. The industrial dyes and the related by-products are carcinogenic and mutagenic, thereby leading to detrimental effects on the ecosystem, including plants, mammals and birds [1 - 5]. Additionally, such effluents result in corrosion and blockages in pipes, uncontrolled eutrophication, bioaccumulation leading to allergies, diarrhoea, immune suppression, haemorrhage, and liver and kidney malfunctioning, dermatitis, DNA damage, neuro-muscular and central nervous system disorder to name a few [1 - 10]. Although there are numerous techniques reported for treating wastewater, novel routes for environmental remediation are on the rise as the existing ones are not adequate to terminate the damage.

Different strategies are being developed for treating the wastewater, and effective and efficient treatment techniques for dye degradation before their discharge into water bodies are the need of the hour.

In the past decade, numerous nanomaterials have been extensively explored for dye degradation applications in treating pollution caused by the release of industrial effluents, particularly xenobiotics [11 - 13]. With the growing demand for noble (Ag, Au, Pd) nanomaterials for different applications, the use of capping agents to achieve accurate control over the size and shape of nanoparticles (NPs) has substantially increased [14 - 18]. Although such nanomaterials are being explored extensively with dielectric and graphene-based hybrids in multidisciplinary applications, the capping agents strongly bind to the metal surface and conceal the intrinsic catalytic activity of the metal [19 - 21]. This chapter is directed towards the use of capping agent free catalysts and to highlight their efficiency in xenobiotic dye degradation. Generally studied dyes are chosen as model substrates for decolorization reactions. The added advantage with this dye decolorization is that the reaction can be easily monitored using a spectrophotometer. By and large, the chapter highlights the role of surfactant-free metal-graphene oxide nanocomposites that are engineered in two different techniques: (i) laser ablation mediated synthesis (LAMS) [22] and (ii) multifunctional soret nano-assemblies (MSNAs) [23 - 26] towards dye degradation and decolorization [27 - 30].

Graphene is a one-atom thick single layer of graphite. It is comprised of sp^2 hybridized carbon atoms with a hexagonal framework [31]. Single layer graphene has unique characteristics which can be briefly listed as (i) a high theoretical specific surface area of 2600 m^2/g, (ii) excellent electrical conductivity of 9.6 x 10^5 S/cm, (iii) great thermal conductance of ~5000 W/Mk and (iv) it supports interesting transport phenomena such as quantum hall effect [32, 33]. Numerous applications are associated with graphene's distinctive properties on account of its multifunctional capabilities [34, 35]. Consequently, graphene is a principal substitute for other forms of carbon as a good catalytic support material. This is on account of its high surface area concomitant with superior conducting properties [36, 37]. This has assured its application in field emission, storage devices, biosensors, super-capacitors, opto-electronics, membrane materials, and electrochemistry [38, 39]. Practical applications of single-layered graphene have become inadequate on account of the cumbersome procedures involved in its development, management/handling and large-scale production [40, 41]. Present synthesis of graphene suffers from scalability, multilayer formation or usage of toxic chemicals during production. In this context, hydrogen exfoliation of graphite oxide offers a compromised approach to acquire enormous quantities of few-layered graphene sheets (BET surface area ~ 430 m^2/g) with notable electrical conductivity [42]. Subsequent treatment of this material by acids has been shown to produce functionalized few-layers graphene (f-HEG). Earlier works have dispersed the f-HEG in aqueous solvents to achieve necessary defect sites for anchoring appropriate functional moieties such as metal/metal oxide nanoparticles [43].

Metal nanoparticles (MNPs), on the other hand, catalyze chemical reactions by an efficient relay of electrons to the substrate under study. The incorporation of graphene's high surface area with MNPs catalyzing power provides unique composite nanomaterials with enhanced catalytic performance. These nano-catalysts not only reduce the reaction time but also pave the way for synthesizing next-generation hybrid graphene-based catalysts. However, the use of surfactants for the stabilization of MNPs modified graphene catalysts has remained a long-standing challenge as it reduces the overall catalytic performance of the substrate [44, 45]. Researchers have synthesized gold graphene composites using bottom-up approaches utilizing different surfactants [46, 47]. In this context, laser ablation of gold strip results in the ablation of pristine surfactant-free AuNPs and also deoxygenation of f-HEG into laser converted graphene (LCG).

The first half of this chapter elaborates on a novel *in-situ* and surfactant-free technique for the synthesis of AuNPs on laser converted graphene (Au-LCG) using laser ablation mediated synthesis (LAMS). The catalyst is utilized in the reduction of Rhodamine B (RB), Methylene blue (MB) dye systems that are widely used in the textile industries. Additionally, Sulforhodamine 101 hydrate (SR) and Fluorescein (FS) that are used as fluorescent staining dyes are also studied. The dye reduction is observed to be kinetically slow in the absence of the catalyst. The results indicate that Au-LCG catalyzes the reduction as well as decolorization of MB at 17000 times faster than the uncatalyzed reaction and 7100 times faster than f-HEG as a catalyst. In the case of RB reduction, the catalyst displayed 1.7 times faster reduction than commercial citrate capped AuNPs of similar size. The added advantage of the decolorization process is that one could follow the kinetics using UV-vis spectrophotometer at ambient conditions (atmospheric temperature and pressure).

The second half of the chapter describes the utility of multifunctional soret nano-assemblies (MSNAs) for dye degradation applications. In this direction, thermal-gradient driven self-assembly of nanoparticles, under adiabatic cooling conditions, is explored for the synthesis of precise nano-assemblies, termed Sorets [23, 28]. The existence of chemical linkers that help to generate the nanoparticle assemblies is a chief contributor towards the intrinsic Ohmic losses and surface functionality. Therefore, an approach to create such nanovoids and nanocavities through non-chemically mediated assembly routes would be attractive for both fundamental physico-chemical explorations and applications in xenobiotic degradation. Despite the capabilities of graphene 2D plasmons to realize distinctive characteristics, their hybrids with plasmonic NPs are not explored hitherto in the growing arena of dye degradation and environmental pollution remediation. A detailed study to probe the reason for the enhanced activity of Au-LCG as well as MSNAs and the mechanistic aspects of the decolorization process has also been studied.

METHODOLOGY

Synthesis of f-HEG and Au-LCG

While the f-HEG was synthesized based on previous reports [43, 48], the Au-LCG was synthesized using the LAMS approach [49]. In brief, 1 mg of f-HEG in 25 mL of millipore® water has been subjected to ultra-sonication to obtain the aqueous solution of f-HEG. A gold metal strip of 2 cm x 1 cm dimension was ultrasonically cleaned in 5 mL piranha solution (caution: handle with care with appropriate laboratory precautions) and placed in a glass beaker containing 3 mL of aqueous solution of f-HEG. A fundamental harmonic at 1064 nm from a high power nanosecond Nd:YAG (Surelite III) was utilized as the source of radiation. A convex lens with a 10 cm focal length was utilized to optically steer and focus the laser pulse energy of 50 mJ with a repetition rate of 10 Hz. These parameters were kept constant for the entire process during the ablation time period. In order to obtain AuNPs (pristine) in Millipore® water, a similar methodology was adopted. To guarantee the release of 0.3 mg of gold into the solution, the weight of the gold strip was monitored after 40 minutes of ablation time. The LAMS procedure is illustrated in Fig. (**1**). Further, the fabrication procedure for MSNAs is detailed in our earlier works [23 - 30].

Characterization

The TEM images of the samples under study were obtained using transmission electron microscopy (TECNAI F–20, S–TWIN, 200 kV). A few drops of the sample were dropped on the Cu grids (SPI supplies, 200 mesh) with carbon coating to prepare the samples for TEM characterization. Indium tin oxide surface has been used as the substrate for performing the FESEM and EDAX analysis (FESEM, Zeiss). Dilor XY triple grating monochromator micro-Raman spectrometer (with a 100X objective) has been used to perform Raman analysis. In this context, samples were excited with n Ar⁺ ion laser with 514.5 nm excitation wavelength and <3 mW. UV-Vis spectrophotometer (Shimadzu 2450 PC) was adopted to characterize the loading of Au on graphene substrate as a function of ablation time.

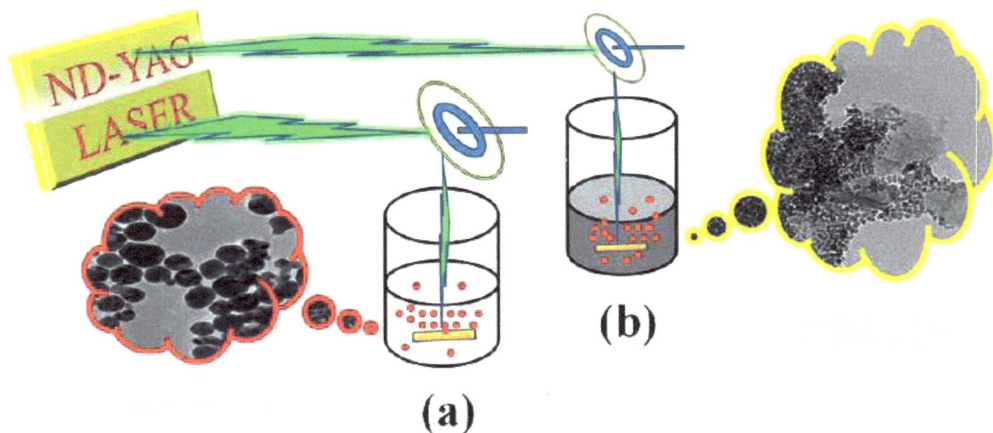

Fig. (1). Schematic illustration of the LAMS of (**a**) AuNPs in water, and (**b**) in an aqueous dispersion of f-HEG. Reprinted with permission from Reference [22], Copyright 2014 Elsevier.

Catalytic Dye Reduction Test

Different dye molecules, including cationic (RB and MB), anionic (FS) and a bipolar dye (SR), were studied to analyze the potential of catalytic reduction of the synthesized nanohybrids. A 10^{-5} M dye in water was obtained upon appropriate dilution of the standard stock solution (10^{-3} M) that was earlier prepared. Aqueous 0.1 M $NaBH_4$, as the reducing agent for the catalytic reaction, was freshly prepared at the time of the experimentation. The reaction was carried out in a quartz cuvette containing 2.5 mL of 10^{-5} M dye, 0.5 mL of 0.1 M of $NaBH_4$ and 50 μL of the catalyst (Au-LCG). Further details of the volume, concentration and absorbance are elaborated in Table **1**.

The same procedure was adopted for the study with AuNPs as well as f-HEG. A Shimadzu 2450 PC UV-Vis spectrophotometer coupled with Julabo F 25 thermostat was utilized to monitor the catalysis. In a few seconds to minutes, total decolorization was noted for all the systems. The controls or blanks were studied with the following three combinations: (i) in the absence of catalyst, that is, only with the use of $NaBH_4$, (ii) in the presence of f-HEG alone and (iii) in the presence of AuNPs alone. The entire study has been repeated with sucrose (viscogen) to systematically investigate whether the reaction was diffusion controlled. Viscogens such as sucrose are known to significantly affect the diffusion controlled reactions, as a result of which we have chosen the same [50]. This was performed by taking 40 mg of sucrose (viscogen) in 10 mL of 10^{-5} M RB dye solution and repeating the reduction procedure to evaluate the reaction rates

[51, 52]. In order to draw conclusive interpretations, the rates were compared with the surfactant capped AuNPs. Citrate capped AuNPs of sizes 20 nm, 80 nm, and 200 nm were purchased from BBI solutions for this purpose. The catalyst was also compared with the plant based AuNPs synthesized using the stem extracts of *Breynia rhamnoides* [52].

Table 1. Tabulated procedure for dye reduction Reprinted with permission from Reference [22], Copyright 2014 Elsevier.

Name of the Dye	Molecular Weight	Absorbance Maximum (nm)	Dye Added		NaBH$_4$ Added		Volume of Au-LCG (µl)
			Molar Conc. (M)	Volume (ml)	Molar Conc. (M)	Volume (ml)	
RB	479.02	553	10^{-3}	2.5	0.1	0.5	50
SR	606.72	588	10^{-3}	2.5	0.1	0.5	50
FS	-	488	10^{-3}	2.5	0.1	0.5	50
MB	319	663	10^{-3}	2.5	0.1	0.5	50

RESULTS AND DISCUSSION

Electron Microscopy, EDAX, XRD, Raman, Zeta Potential Studies and UV-Vis Spectroscopy of f-HEG and Au-LCG

The synthesized samples have been examined by electron microscopy and results are presented in Fig. (**2**). Fig. (**2a**) shows a TEM image of f-HEG where a few layers are observed. The TEM image of AuNPs presenting a spherical morphology is shown in Fig. (**2b**). The inset of the same Fig. shows the particle size distribution, which indicates the particles to be in the range of 20-25 nm. The TEM image of Au-LCG is given in Fig. (**2c**), and the high-resolution transmission electron microscopy (HRTEM) image of the same is shown in Fig. (**2d**), along with the characteristic d-spacing. The dark field TEM image of Au-LCG under high contrast is shown in Fig. (**2e**). The FESEM image of Au-LCG is given in Fig. (**2f**), further supporting the claim about the morphology. The results of the EDAX elemental analysis are given in Table **2**. From the EDAX analysis, it was observed that the amount of Au, excluding the ITO background, is 52.65%.

Fig. (**3a**) and Fig. (**3b**) present the Raman spectra for f-HEG and Au-LCG, respectively. It is observed that the Raman features of the f-HEG samples are similar to that of the Raman features generally obtained by graphene, which is produced by the exfoliation of graphite oxide (GO). This is on account of the defective and few-layered nature that has also been seen in the TEM images discussed in an earlier section. Multiple defects activated bands were observed upon the deconvolution of the 2D band occurring at 2680 cm^{-1}. It has been reported that subjecting conventional graphene to Ar ion bombardment [53] results in a similar G' band profile. Further, a blue shift is observed for the G and the 2D bands by 4 and 6 cm^{-1}. This is attributed to the weak charge transfer from laser converted graphene (LCG) to AuNPs in the Au-LCG composite. It may be hypothesized that the presence of defects in the sample supports the tethering of AuNPs. This would, in turn, assist in the enhanced catalytic activity of the composite hybrid [54].

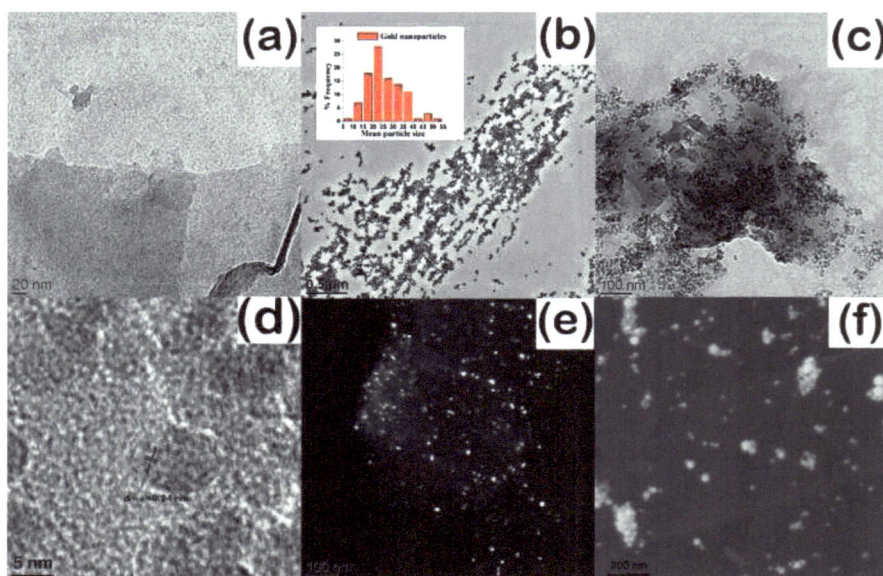

Fig. (2). (a) TEM image of f-HEG (b) TEM image of pristine AuNPs inset shows their size distribution (c) TEM image of Au-LCG (d) HRTEM image of Au-LCG shows the interplanar spacing of AuNPs (e) dark field TEM image of Au-LCG under high contrast and (f) FESEM image of Au-LCG. Reprinted with permission from Reference [22], Copyright 2014 Elsevier.

Table 2. EDAX Elemental analysis results for Au-LCG. Reprinted with permission from Reference [22], Copyright 2014 Elsevier.

Elements	Atm %	Wt %
C	17.59	8.15
O	72.72	44.88
In	6.76	29.97
Sn	1.73	7.94
Au	1.19	9.06
Total	100	100

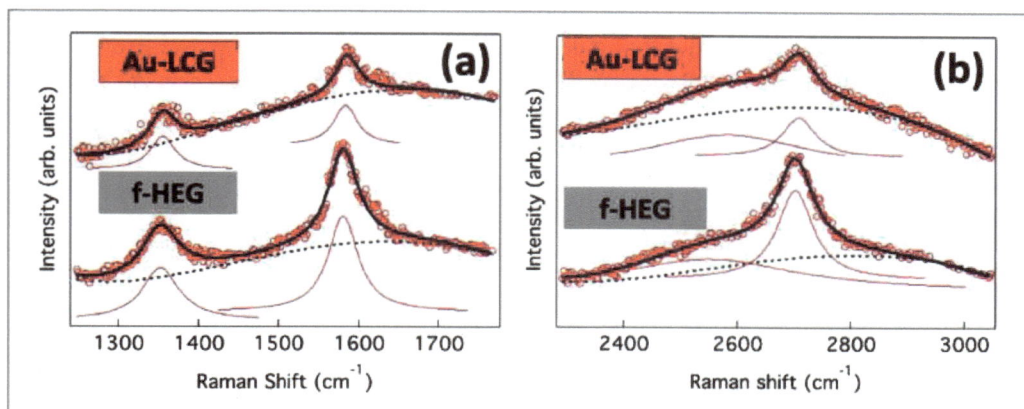

Fig. (3). Raman spectra of f-HEG and Au-LCG. The solid black lines show the fit. The deconvoluted peaks are shown below the obtained spectra along with the background (dotted lines). Reprinted with permission from Reference [22], Copyright 2014 Elsevier.

The stability of the aqueous dispersion of synthesized f-HEG was evaluated by storing the sample for an extended period of time of 92 days, as shown in Fig. (**4**). It was observed that the sample is stable for more than 3 months, emphasizing the high stability of the composite hybrid. In order to confirm whether the f-HEG present in the dispersion is also affected in the laser ablation process of gold, the laser irradiation of f-HEG alone has been studied. With this, it was noted that the latter process resulted in the deoxygenation of f-HEG. This is confirmed by the UV-vis spectra captured in Fig. (**5a**), where one can note the disappearance of the absorbance peak at 233 nm, with a concomitant increase in the intensity of the peak at 265 nm. This specifies the restoration of sp^2 carbons by deoxygenation of f-HEG upon laser irradiation at a 1064 nm laser pulse. Since f-HEG endows itself with less oxygen content, extensive wrinkly nature, superior C/O ratio and electronic conductance, this material has been chosen in this case [28]. In the case of f-HEG (in comparison to GO), it is less difficult to deoxygenate f-HEG and simultaneously load AuNPs. Needless to say, the π-π stacking occurs between the graphene parent structure and the dye molecules, thereby re-emphasizing the need for graphene in the sample. In addition to this, the intensity at 520 nm, which is characteristic to that of localized surface plasmon resonance (LSPR) of AuNPs, increases upon increase in the laser ablation time (Fig. **5b**).

The distinction between the f-HEG and Au-LCG is evidently seen in the XRD spectra (Fig. **6**). A sharp peak in the XRD spectra is observed at 26.7° corresponding to the (002) of the hexagonal framework of graphite. This also corresponds to the d-spacing of 3.4 Å. Also, the peak seen at 10.9° is attributed to the intercalated graphitic sheets with a d-spacing of 8.4 Å [55]. Further, for the f-

HEG sample, a broad peak is observed from 14° to 30° (with a corresponding d-spacing of 3.7 Å). This is due to the reduction in the long-range graphitic order on account of the hydrogen exfoliation. Additionally, the 2θ shift towards the GO at 10.9° observed is attributed to the oxygenation of the graphene parent structure. The LCG and Au-LCG formation occurred on account of the synchronized deoxygenation of f-HEG and the loading of AuNPs. It has been observed that the Au-LCG did not possess the 10.9° GO peak as well as the broad f-HEG peak that was observed earlier. The higher similarity of the LCG (acquired by laser irradiation routes) to that of the graphene samples is attributed to the absence of peaks in the region from 0° to 30°, as shown in Fig. (**6a**) [54, 56]. This is due to the reduction process that led to restoration of the sp^2 carbons in the graphene parent structure. From Fig. (**6b**), we see that the Au-LCG presented many peaks in the region from 30° to 90°, in contrast to f-HEG. The characteristic crystalline peaks at 38.1°, 44.3°, 64.7°, and 77.7° correspond to (111), (200), (220) and (311) planes of AuNPs, respectively. The zeta potential measurements presented interesting results, where it was observed that the pristine AuNPs and Au-LCG presented -40.0 mV and -48.1 mV thereby confirming the high stability of the composites synthesized as compared to pristine AuNPs. In accordance with the earlier work [57], subjecting the AuNPs to LAMS at pH more than 5.8 resulted in the formation of Au-O⁻ species, which render a negative charge to the overall sample under study. We anticipate the formation of similar species in our work as the pH has been maintained at 7. It is worth mentioning that on account of decreased (more negative) zeta potential of the composite hybrid (as compared to the pristine AuNPs), the synthesized samples retain long-term stability. These composite hybrids were further considered as catalysts for dye decolorization and reduction analysis.

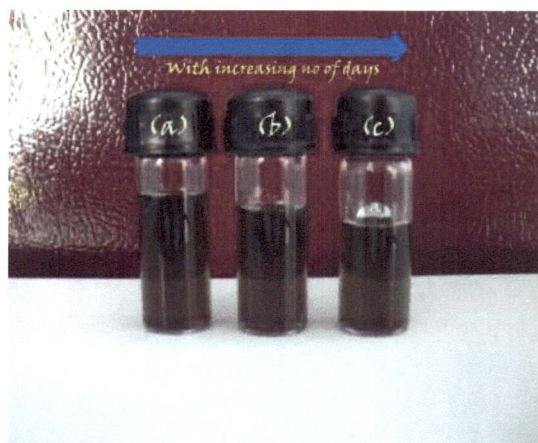

Fig. (4). Aqueous dispersion of f-HEG without any marked difference for (**a**) 1 day (**b**) 35 days and (**c**) 92 days. Reprinted with permission from Reference [12], Copyright 2014 Elsevier.

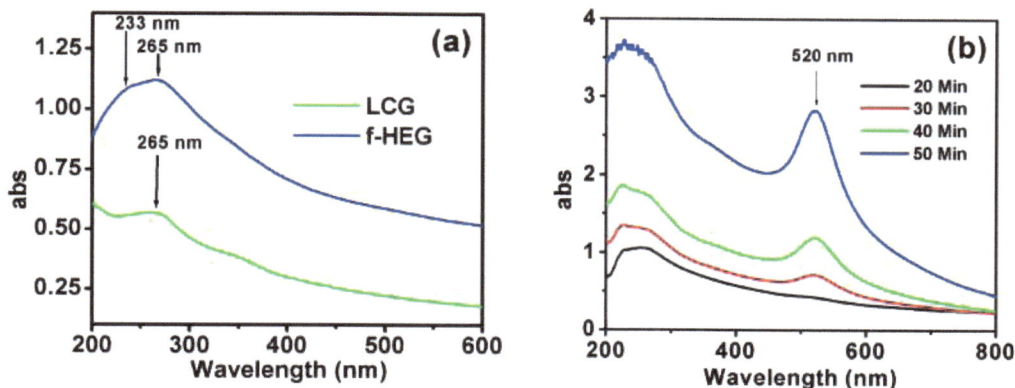

Fig. (5). UV-Vis spectra overlay of (**a**) f-HEG and LCG (**b**) Au-LCG at different ablation times. Reprinted with permission from Reference [22], Copyright 2014 Elsevier.

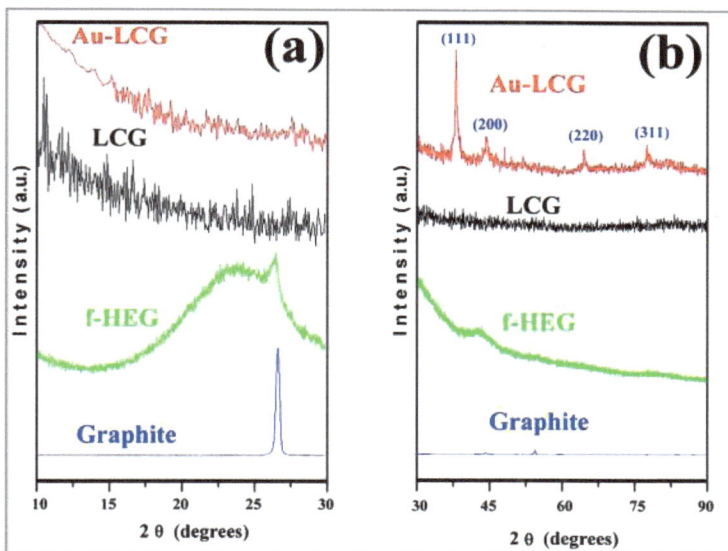

Fig. (6). XRD overlay of Graphite, f-HEG, LCG and Au-LCG. Reprinted with permission from Reference [22], Copyright 2014 Elsevier.

Analysis of Catalytic Activity of Au-LCG in Dye Reductions

This experiment supports better comprehension of the potential of novel nanomaterials such as Au-LCG toward dye degradation reaction. NaBH$_4$ has been used as the reducing agent which involves electrons as charge carriers during the catalysis [58]. Earlier work demonstrates the utility of novel systems such as Ir nanoneedles to augment the electron flow from the reducing agent to the dye molecules to hasten the entire reduction process [58]. In this perspective, the principal aim of the current work is to understand the potential role of Au-LCG to act as a catalyst in the electron relay from the electron source ($NaBH_4$) to the electron sink (dye molecules).

The reaction of the generation of reduced species was governed by pseudo-first order reaction kinetics. This is confirmed by the spectroscopic details given in Table **3** and Fig. (**7**). In order to understand the reaction kinetics without the reducing agent, the reaction was separately carried out in the absence of borohydride. In this scenario, no reduction was observed, emphasizing the need for the reducing agent or the electron source in the reaction mixture. Further, the reaction was also carried out in the presence of borohydride reducing agent and the dye system alone (*i.e.*, without any catalyst). In this case, as well, a considerable change in the rate of the reaction was not observed, as detailed in Table **3**.

Table 3. Kinetic details of dye reduction with Au-LCG and various experimental controls. Reprinted with permission from Reference [22], Copyright 2014 Elsevier.

Dyes	AuNPs		Au-LCG		Dye+NaBH$_4$	f-HEG	Ir Nanoneedles [28]	
-	k_{AuNPs} (s^{-1})	time (s)	k_{Au-LCG} (s^{-1})	time (s)	$k_{Dye+NaBH4}$ (s^{-1})	k_{f-HEG} (s^{-1})	k_{Ir} (s^{-1})	time (s)
RB	8.6×10^{-2}	86	2.6×10^{-1}	69	1.5×10^{-3}	1.9×10^{-3}	1.9×10^{-3}	1440
SR	1.6×10^{-2}	150	1.1×10^{-1}	81	1.5×10^{-4}	4.7×10^{-5}	-	-
FS	1.0×10^{-3}	2100	2.3×10^{-3}	1032	2.0×10^{-4}	7.0×10^{-5}	-	-
MB	7.5×10^{-1}	2	6.4×10^{-1}	8	3.7×10^{-5}	9.0×10^{-5}	6.8×10^{-4}	3600

Before the reduction of the dyes, a lag phase was observed during the kinetic studies of the Au-LCG. This is due to the formation of the activated catalyst complex (ACC). The diffusion-controlled adsorption of dyes and borohydride ions onto the Au-LCG surface resulted in lag time. The schematic representation given in Fig. (**8**) demonstrates the view of the ACC catalyst as AuNPs (on LCG) covered by borohydride [58] which is in the immediate vicinity of the dye system. Upon the development of the ACC catalyst, the dyes in the close vicinity are reduced. The overall lag phase or the time taken for the ACC generation is depended on parameters such as diffusion, sorption and orientation of the BH_4^- and the dye, with respect to the Au-LCG.

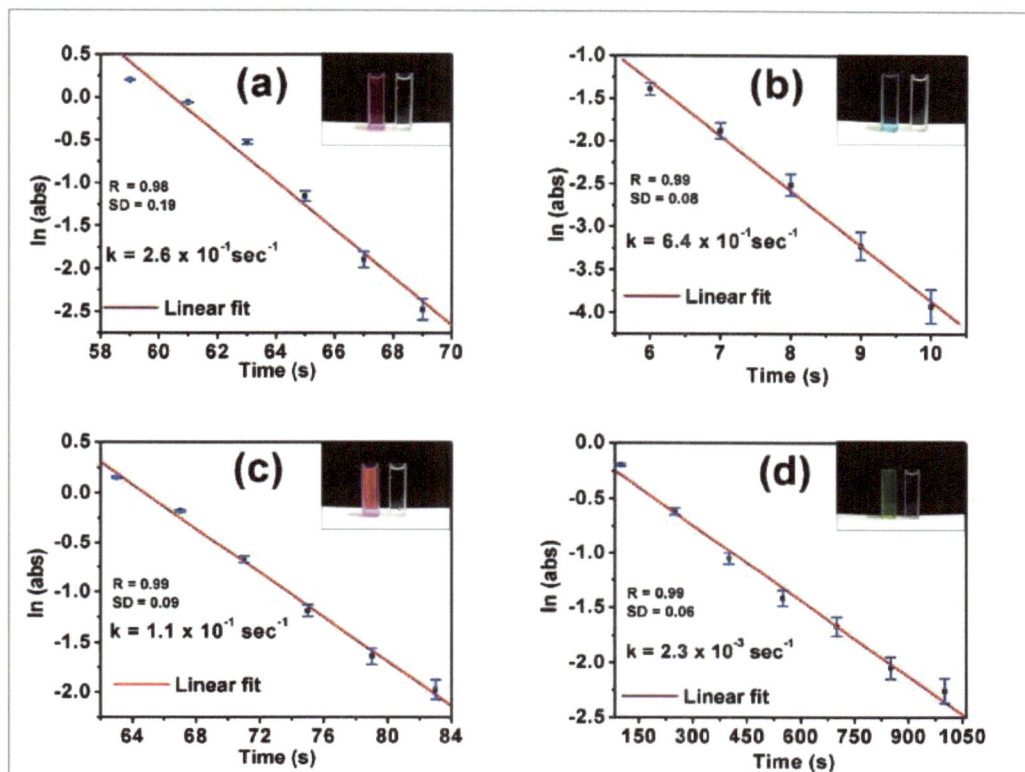

Fig. (7). Plots of ln (abs) *vs.* time for reduction of (**a**) RB, (**b**) MB, (**c**) SR, (**d**) FS using Au-LCG composite as catalyst. Reprinted with permission from Reference [22], Copyright 2014 Elsevier.

Furthermore, it is important to consider the chances of the dissolved oxygen in the solvent to compete with the dyes for the borohydride ion formation. In order to completely understand this phenomenon that the lag phase is due to the diffusion-controlled adsorption of the reactants over the catalyst, an additional set of experiments was performed using viscogens. Viscogens such as sucrose are known to significantly affect diffusion-controlled reactions, as a result of which we have chosen the same. Upon the addition of the sucrose viscogen, changes were introduced in the reaction rates and the lag time. This confirms that the lag time is principally on account of the diffusion-controlled adsorption of the reactants over the catalyst as it can be seen in Fig. (**9**). The process has been cross-validated using ethanol as another viscogen and the results are presented in Fig. (**10**). The reaction to be diffusion controlled has been verified by the perseverance of the lag time of the reaction in an inert atmosphere which has been achieved by purging the nitrogen.

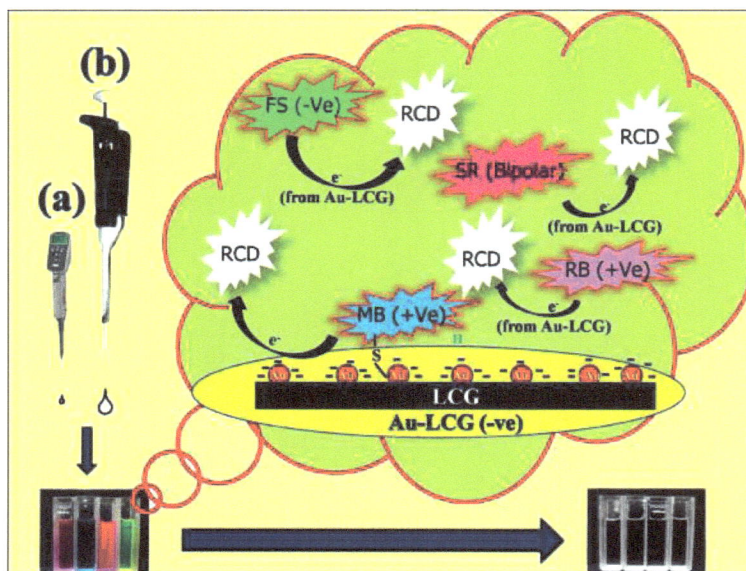

Fig. (8). The process of dye reduction on addition of (**a**) Au-LCG (catalyst) and (**b**) NaBH$_4$ (reductant). The callout depicts schematically the mechanism of dye reduction on Au-LCG to RCD (reduced colorless dye). Reprinted with permission from Reference [22], Copyright 2014 Elsevier.

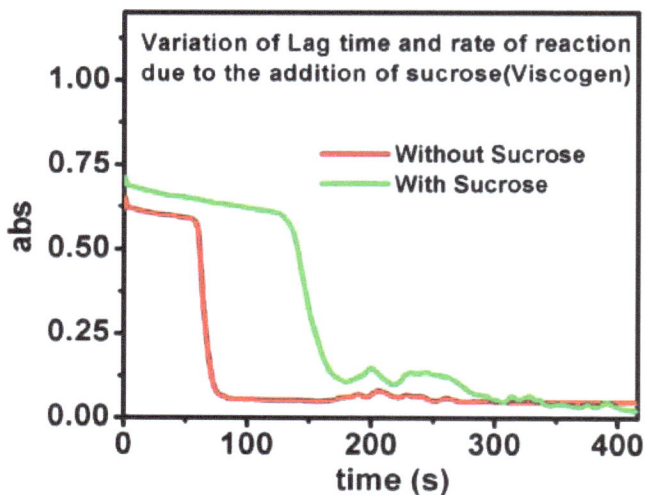

Fig. (9). A UV-Vis kinetic study of RB reduction in the presence and absence of sucrose as a viscogen. Reprinted with permission from Reference [22], Copyright 2014 Elsevier.

Table **3** gives a comprehensive analysis of the rates of the reaction obtained for all the dye systems which was studied using simple UV-Vis spectrophotometer. A change in the wavelength maxima of the absorbance spectra has been observed with change in the concentration of the dye. From these analyses, it has been concluded that the AuNPs surface is better as compared to the Ir surface for electron relay. This is on account of the augmentation in the metal to ligand charge transfer mechanism [59]. Over and above this, it must be noted that the substantially high rates obtained with the use of Au-LCG (as compared to AuNPs) is due to the crucial participation of LCG in the composite. These aspects, such as approach, orientation, sorption and distance of the dye in the ACC system, are directed by the LCG sample. The reduction rates of the reaction for cationic (RB and MB), anionic (FS) and bipolar (SR) dye systems shown in Table **3**, gives more clarity to this understanding.

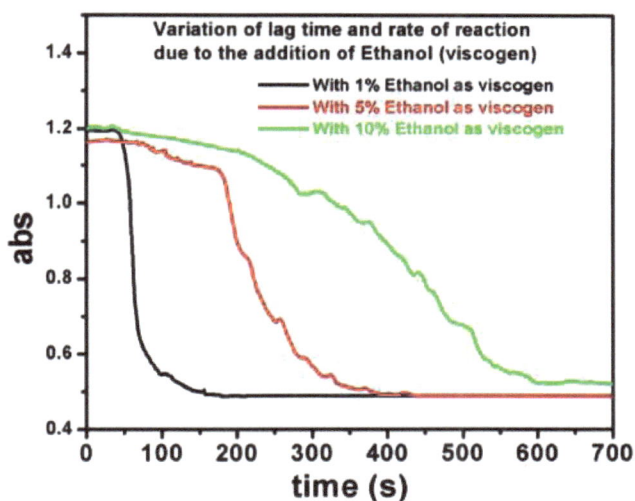

Fig. (10). A UV-Vis kinetic study of RB reduction in the presence of different Ethanol concentrations as a viscogen. Reprinted with permission from Reference [22], Copyright 2014 Elsevier.

Important information about the surface charge and basic properties can be obtained from the molecular structure of the dye molecule, as shown in Fig. (**11**). It can be seen that, in line with our earlier understanding, the Au-LCG (negatively charged) demonstrates a high affinity toward cationic dues. This is followed by the bipolar dyes and finally by the anionic dyes as shown in Table **3**. From this understanding, it can be observed that in the presence of Au-LCG, the reduction rates of RB, SR and FS dye systems are ~3, 6.9 and 2.3 times faster as compared to pristine AuNPs. Similarly, the comparison of the Au-LCG and HEG with the uncatalyzed reaction is tabulated in Table **4**. The higher reduction rates of the Au-LCG are on account of the π-π interaction between the LCG and the dyes under consideration. Here, a van der Waals complex formation is expected on account of the weak electrostatic interaction between the ground state of the dye and the LCG [60]. As a result of this close interaction, the dye molecules are drawn towards the Au-LCG, which in turn propels the reduction rate.

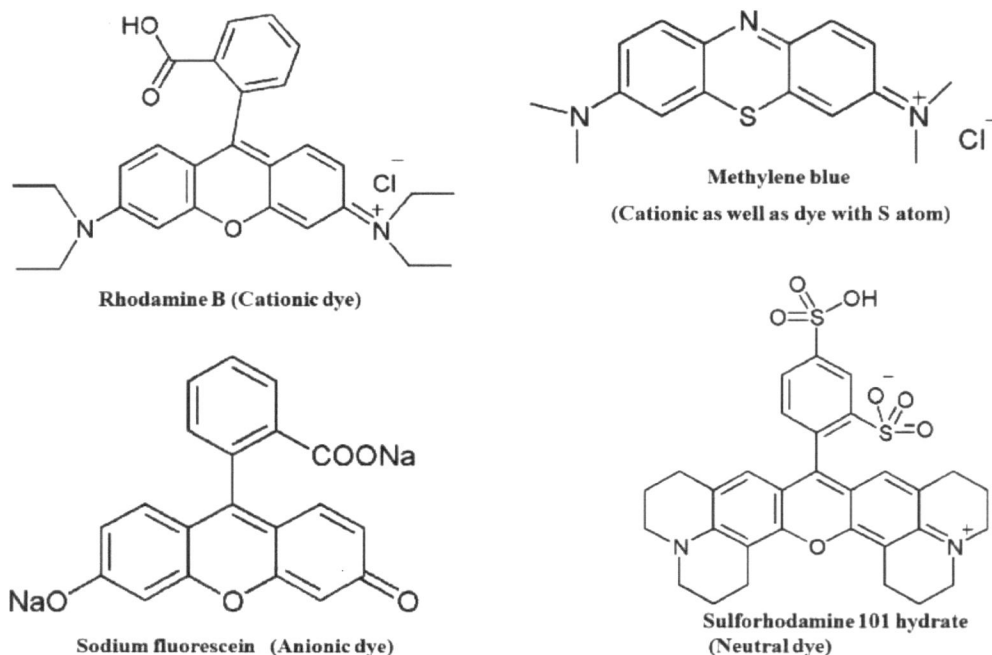

Rhodamine B (Cationic dye)

Methylene blue
(Cationic as well as dye with S atom)

Sodium fluorescein　(Anionic dye)

Sulforhodamine 101 hydrate
(Neutral dye)

Fig. (11). Molecular structures of different dyes studied in this work. Reprinted with permission from Reference [22], Copyright 2014 Elsevier.

Table 4. Number of folds of increase in reaction rate for all the 4 dyes with Au-LCG as a catalyst compared to the uncatalyzed reaction, f-HEG and Pristine AuNPs. Reprinted with permission from Reference [22], Copyright 2014 Elsevier.

Dye	Uncatalyzed Reaction	f-HEG	Pristine AuNPs
RB	173.3	136.8	3
MB	17297.3	7111.1	-
SR	733.3	2340.4	6.9
FS	11.5	32.9	2.3

It is important to discuss an exception that was observed with the use of MB. Here the reaction rate is observed to be 1.2 times faster with AuNPs as compared to that with the Au-LCG composite. This is intuitive as the MB dye possesses a sulfur atom as well as the nitrogen atom that aid in covalently anchoring the molecule on the gold surface [61]. It was observed that the reduction rates were 140 and 950 times faster for RB and MB dyes, respectively, in comparison with the Ir nanoneedles [58]. Additionally, it is to be noted from Table **5** that with an increase in the Au-LCG concentration, the reaction rates increased, further confirming the ACC formation. Moreover, the reaction was studied in dark conditions to obtain a comprehensive understanding of the efficiency of the process. It was observed that the dye reduction by Au-LCG was achieved even in dark conditions, unlike the earlier reported photo-degradation of dye systems [62].

Table 5. Concentration dependent reduction of different dye systems using Au-LCG composite. Reprinted with permission from Reference [22], Copyright 2014 Elsevier.

Volume of Au-LCG (µL)	Time Taken (Seconds) for Dye Reduction			
	RB	**MB**	**SR**	**FS**
10	696	22	867	2100 (Partial reduction)
50	69	8	81	1032
100	39	-	51	803

The entire process of dye reduction was repeated with commercially available citrate capped AuNPs of 20 nm, 80 nm, and 200 nm as well as AuNPs synthesized from the stem extract of *Breynia rhamnoides* [52]. The representative plots for the reaction rates obtained with the commercial and plant-based nanoparticles for the reduction of RB as shown in Fig. (**12**). The catalytic reduction rate was better in Au-LCG, *i.e.*, 1.7 times more than commercial citrate capped AuNPs (20 nm). The catalytic efficiencies of Au-LCG as compared to different AuNPs as well as different controls are shown in Fig. (**13**).

MSNAs for Dye Degradation Application

The nanoparticle assemblies, termed Sorets colloids (SCs), were synthesized using adiabatic cooling methodology that has been reported earlier [23 - 30]. The linker-free assembly of NPs in the SCs generates nanovoids and nanocavities that are critical to achieve efficient interparticle plasmonic coupling by minimizing the intrinsic damping losses and thereby driving highly intense hotspots. The MSNA fabricated, thus, are characterized through various spectroscopic and microscopic techniques to derive a nanostructure-property correlation, as detailed in this section. We have termed it multifunctional on account of its ability to perform excellently in different research fields, including metal-enhanced fluorescence, biosensors and dye degradation.

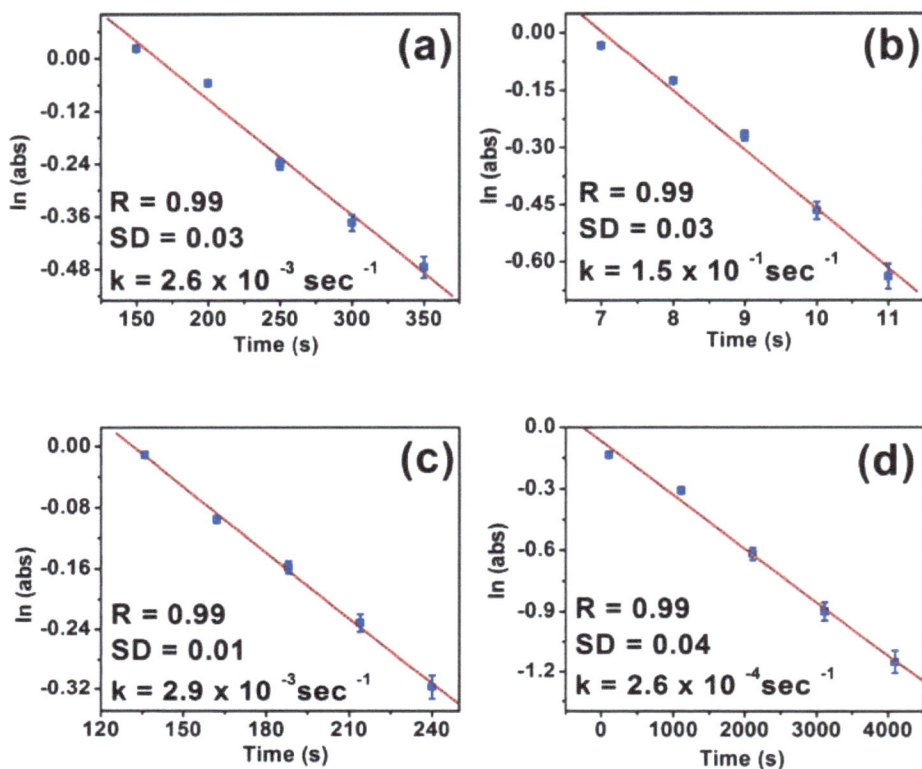

Fig. (12). Plots of ln (abs) *vs.* time for RB dye reduction using (**a**) AuNPs synthesized from stem extract of *Breynia rhamnoides* (Plant based AuNPs) and commercial citrate capped AuNPs purchased from BBI solutions of sizes (**b**) 20 nm, (**c**) 80 nm, (**d**) 200 nm. Reprinted with permission from Reference [22], Copyright 2014 Elsevier.

The multiple TEM images of the AgGOSC are shown in Fig. (**14**). We notice that the GO sheets (that are micron size long) are inter-twined with the silver soret colloids (AgSCs) on account of their sub-nano thickness. The nano-folding in the adjacently located GO sheets assists in viewing the ~1 nm thick GO sheets, in spite of them being highly transparent to the incident electron beam in the TEM experimentation [20, 21, 63]. The TEM characterization aids in clear visualization of nanovoids and nano-crevices comprising the GO flakes in and around the nanohybrid. Fig. (**14e**) demonstrates the characteristic d-spacing of parent graphitic framework in the GO flakes that are part of AgGOSC [20, 21]. On the contrary, the AgNPs appear dark in the TEM imaging compared to the GO, as AgNPs are less transparent to the incoming electron beam. Fig. (**14f**) demonstrates the d-spacing characteristic to that of Ag in the AgGOSC nanohybrid.

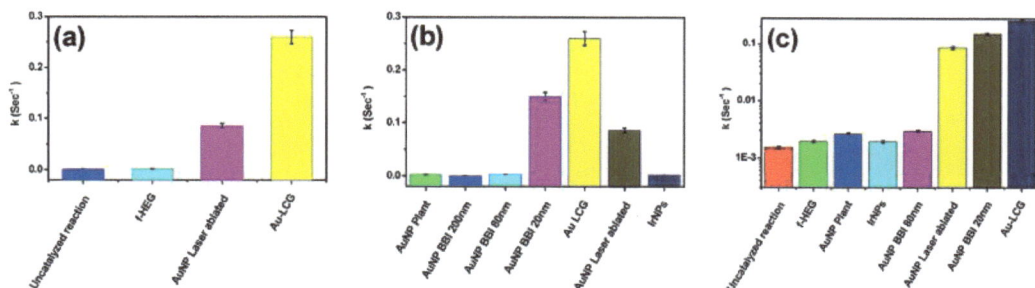

Fig. (13). (**a**) and (**b**) Histograms depicting catalytic efficiencies of Au-LCG over various controls and different AuNP catalysts, (**c**) Different controls and catalysts vs reaction rate in exponential scale. Reprinted with permission from Reference [22], Copyright 2014 Elsevier.

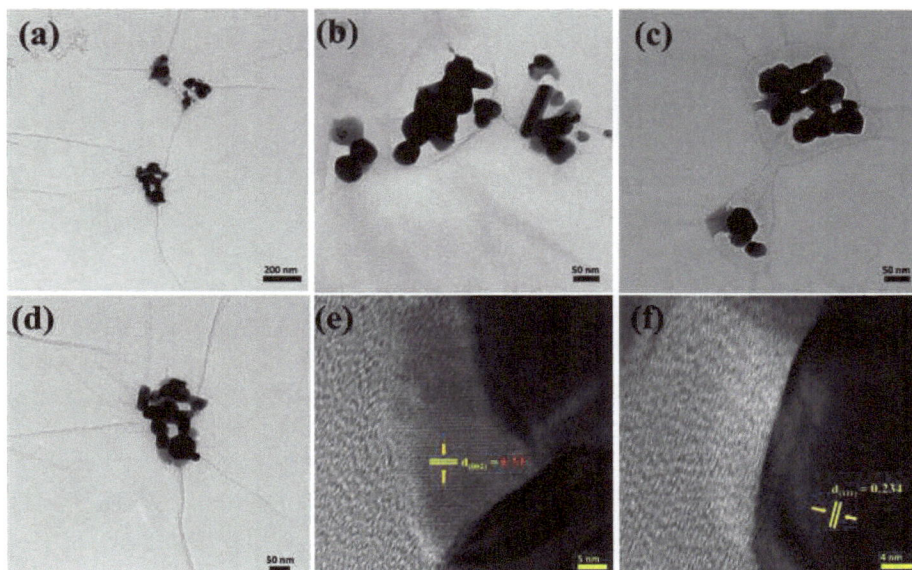

Fig. (14). (**a**), (**b**), (**c**) and (**d**) are multiple TEM images of AgGOSC. (**e**) and (**f**) are the HRTEM images of AgGOSC indicating the characteristic d-spacing of both GO and AgNPs. Reprinted with permission from Reference [28], Copyright 2021 Elsevier.

In order to validate the understanding of the nanohybrid formation in Fig. (**14**), the SAED patterns are captured in two different regions: (i) at the junction between the Ag and GO (Fig. **15a**), (ii) in the centre region of GO (Fig. **15b**) that is part of the AgGOSC nanohybrid. From 15a, one can clearly observe that the bright spots corresponding to that of Ag are distributed on a relatively less bright background of six-fold hexagonal array of diffraction bright spots. This confirms the overall hybrid nature of the soret nano-assembly. In the case of Fig. (**15b**), we observe only a six-fold hexagonal array of diffraction bright spots, confirming the predominant presence of GO in the sample. Furthermore, line profile studies were performed to understand the monolayer characteristic of the GO sample under study. Fig. (**15c**) indicates that the relative intensity of the outer rings ($2\bar{1}\bar{1}0$-type) is lesser compared to that of inner rings ($1\bar{1}00$-type), thereby confirming the sample to be present predominantly as a monolayer, in accordance with the earlier works [20, 21, 25, 63]. The Miller−Bravais *hkil* notation is used for the notation in these Figures.

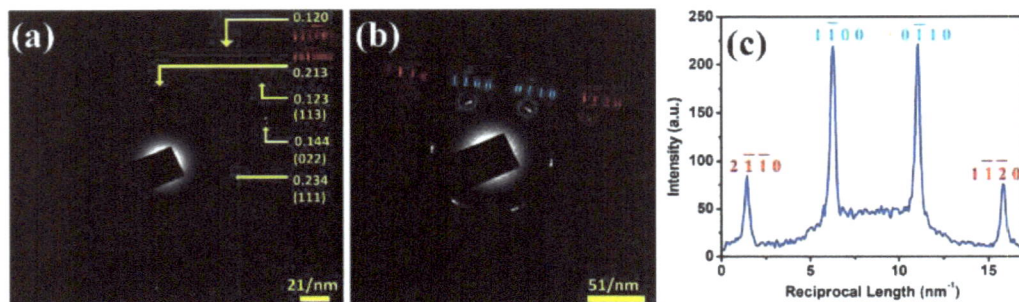

Fig. (15). (**a**) SAED at the junction of Ag and GO in this hybrid soret shown in Fig. 14(**a**). (**b**) SAED at the center region of showing bright spots labelled with Miller−Bravais indices. (**c**) Intensity across the line-profile as presented in (**b**) including inner and outer diffraction dots. Reprinted with permission from Reference [18], Copyright 2021 Elsevier.

Furthermore, an extensive UV-Vis absorbance analysis has been performed to understand the inter-plasmonic coupling and LSPR of the MSNAs. The absorbance spectra of MSNA along with their individual precursor components are shown in Fig. (**16**). The simple mixture of the precursor components used in the soret synthesis displayed different spectra compared to the MSNAs (as it is evident from Fig. (**16b**), blue and dark blue spectra). The spectral change in the formation of AgSC using AgNPs, also reported in our earlier works [24 - 28], is shown in Fig. (**16a**). This distinction is clear on account of the spectral broadening observed at 400 nm to 500 nm in the case of MSNAs. Additionally, longitudinal absorbance in MSNA is on account of the anisotropic clustering effect that results in inter-plasmonic coupling between the NPs of single MSNA. Yet another interesting characteristic observed using GO is the negligible absorbance in the UV-Vis range as compared to that of MSNAs. On the other hand, the MSNAs present both the transverse and longitudinal LSPR, thereby confirming the formation of nanohybrids and differentiating them from a simple physical mixture of individual components (showing only transverse modes of pristine unhybridized AgNPs).

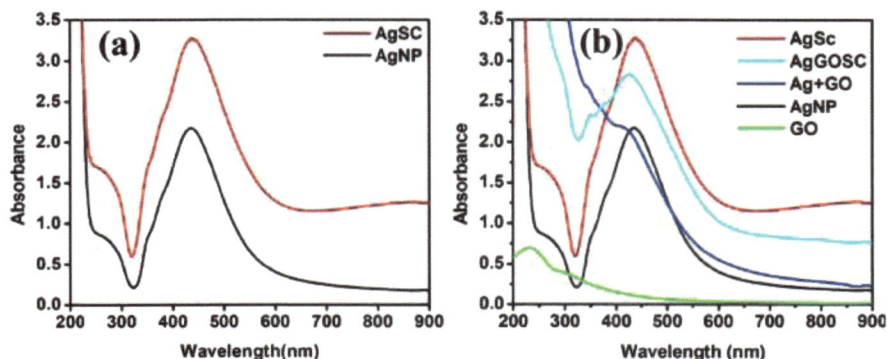

Fig. (16). UV-Vis absorbance profile of (**a**) AgNPs and AgSC, (**b**) AgGOSC along with their starting materials: AgSC, AgNP, GO and the physical mixtures of AgSC and GO. Reprinted with permission from Reference [18], Copyright 2021 Elsevier.

Plasmonic nanomaterials are extensively explored in interdisciplinary research on account of their unique opto-electronic properties [64 - 69]. Earlier reports [22, 28] emphasize the utility of plasmonic NPs and graphene oxide hybrid for decolourization and degradation of harmful organic pollutants (dye molecules) such as rhodamine and methylene blue. In this background, we have examined the efficiency of MSNAs towards dye degradation and the results are interesting. All the experiments were performed as reported in an earlier work [22]. On account of the effectual interaction between the π-plasmons of GO and that of organic dyes, the dynamics in the AgGOSC was substantial due to added contribution from the LSPR of AgNPs. In the case of methylene blue, 63 times faster dye degradation rate in AgGOSC ($k = 3.77 \times 10^{-2}$) as compared to GO solution ($k = 5.98 \times 10^{-4}$) was observed. Blank measurements were carried out with the pristine AgNPs and AgGOSC to obtain conclusive comparisons. This was also ~4 times faster than the degradation achieved by pristine AgNPs ($k = 8.53 \times 10^{-3}$) and AgSC ($k = 9.08 \times 10^{-3}$) samples. In the case of rhodamine 6G, the rate of degradation was 6 times faster in AgGOSC ($k = 2.43 \times 10^{-2}$) compared to that in GO solution ($k = 4.0 \times 10^{-3}$), and ~2 times faster than that in AgNPs ($k = 1.21 \times 10^{-2}$) and AgSC ($k = 1.37 \times 10^{-2}$) solutions.

The dye degradation details for rhodamine 6G and methylene blue along with their positive and negative controls are presented in Fig. (**17** and **18**), and the R values are also shown for the respective linear fits. Further, it is important to note that the rate of reduction for the dye systems in this work is 12 and 55 times faster as compared to that with Ir nanoneedles [70]. This augmented dye reduction rate is due to the close interaction and π-π stacking between GO of the MSNA and dye π conjugated dye molecules. This enhances the electron relay in the nanoassembly of the AgGOSC, thereby presenting MSNAs as effective functional materials for applications in the environmental monitoring of xenobiotics.

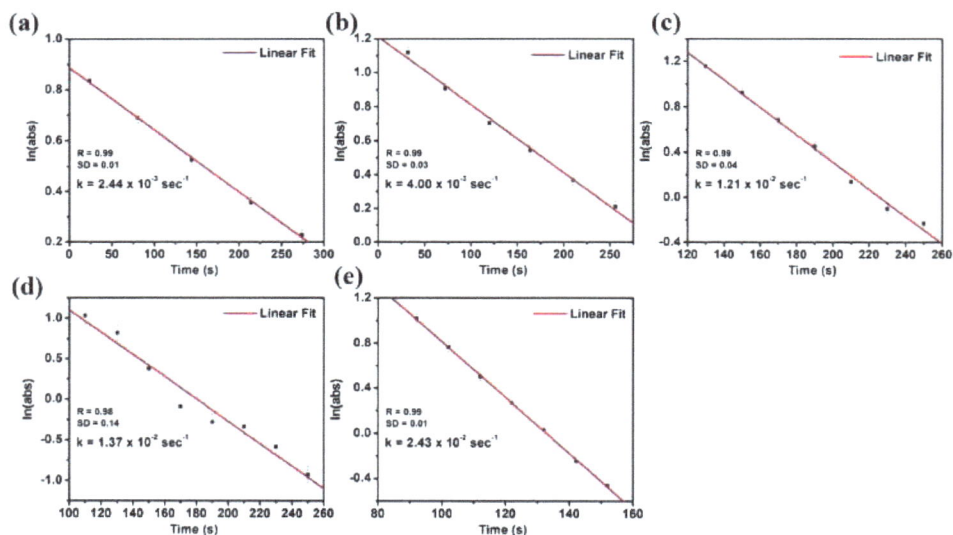

Fig. (17). UV-Vis kinetics of reduction of Rhodamine 6G (λ_{max} = 526 nm) by NaBH$_4$ with (**a**) no catalyst, (**b**) GO as catalyst, (**c**) AgNP as a catalyst, (**d**) AgSC as a catalyst and (**e**) AgGOSC as a catalyst. Reprinted with permission from Reference [28], Copyright 2021 Elsevier.

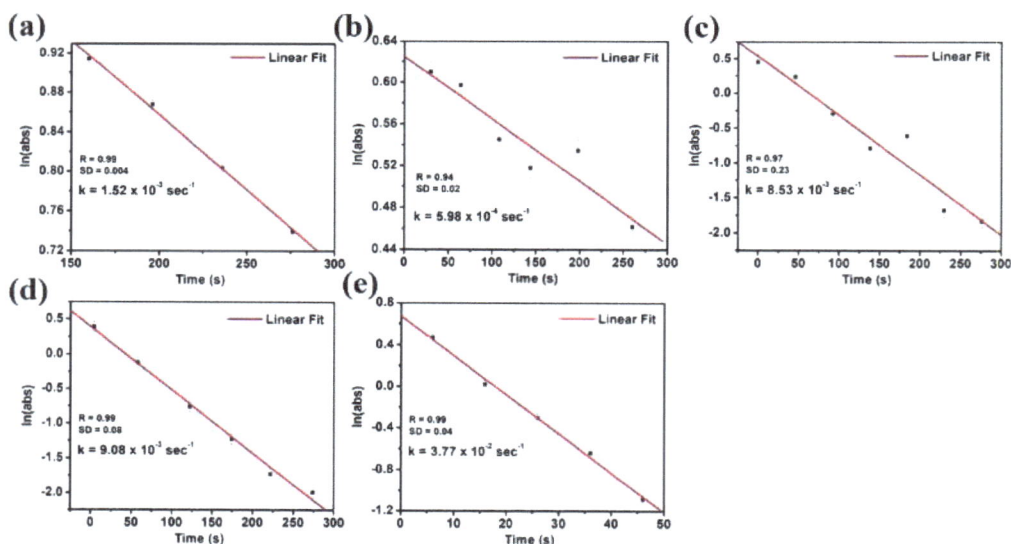

Fig. (18). UV-Vis kinetics of reduction of methylene blue (λ_{max} = 664 nm by NaBH$_4$ with (**a**) no catalyst, (**b**) GO as a catalyst, (**c**) AgNP as a catalyst, (**d**) AgSC as a catalyst and (**e**) AgGOSC as a catalyst. Reprinted with permission from Reference [28], Copyright 2021 Elsevier.

Significant advancement has been made in the research pertaining to the treatment and management of dye wastewater [1 - 3]. Broadly, the wastewater treatment strategies can be classified into four categories, namely, (i) preliminary treatment, including neutralization and equalization, (ii) primary treatment, including sedimentation, chemical coagulation, flocculation, screening, and floatation, (iii) secondary treatment including chemical/physical separation and biological oxidation technique (anaerobic or aerobic) used to reduce organic compounds, and (iv) tertiary treatment of industrial wastewater treatment that specifically includes the hazardous dye degradation techniques [2]. While this is one strategy of classifying the methods involved in general wastewater treatment, the classification in particular for dye hazardous removal is broadly classified into the following: (i) physical, (ii) chemical and (iii) biological treatments and (iv) nanoparticle-based method [3]. It is instructive to provide a brief account of these techniques to present the discussion of this chapter in perspective.

Physical methods mainly include coagulation and flocculation, which is generally effective for dispersal and sulphur-based dyes. This methodology was reported by Beluci *et al.* in combination with ultrafiltration in TiO_2-modified membranes [3, 71]. Adsorption-based method is another efficient strategy where activated carbon (for acidic, cationic & mordant dye removal) with and without silver nanoparticle loading has been explored in the past [72 - 74]. For developing cost-effective adsorbents, recently peat, fly ash, bentonite clay, polymeric resins, ion exchangers, maize and wheat straws, to name a few, are being utilized. However, on account of high sludge production as well as disposal and regeneration issues, some of these adsorbents are less explored [73]. In terms of chemical recovery and reuse of water, certain other physical methods, such as ultrafiltration, reverse osmosis and nano-filtration are effective [3, 73]. High cost, membrane fouling and secondary waste generation are some of the limitations of these techniques. Yet another physical method that is popular is the ion exchange method, where the ion exchange resins are used to treat dye (anionic and cationic) containing effluents [3, 73].

Chemical methods generally adopt oxidizing agents such as hydrogen peroxide, ozone and permanganate have been widely used to chemically distrust or decompose different dyes. Chlorinated hydrocarbons, pesticides & different aromatic hydrocarbons can be treated with the use of ozone [3, 73], in spite of the high cost of ozone. Photochemical and photocatalytic oxidation processes are used in synergy with ozonation, where the oxidizing agents such as TiO_2, ZnO, iron and H_2O_2 *etc.*, in the presence of radiation, destruct the harmful dyes on account of the generation of hydroxide (OH·) radicals [2, 3, 73]. In this methodology, the rate of dye degradation (that finally results in CO_2 and water) depends on the intensity of UV used, the chemical structure of the dye and also the composition of the entire bath. Good decolourization is achieved with a fenton reaction that is comparatively cheap. Although this can be applied for treating soluble and insoluble dyes, high sludge production on account of flocculation restricts its use [3, 73]. Further, electrochemical techniques are also used for treating organic dyes with effective results and less undesirable by-products. However, the need for and the use of electricity typically adds to the expenses of the methodology [3, 73].

Biological methods with the use of bacterial and fungal systems are widely used on account of the advantages such as eco-friendly, cost-effective and non-toxic with the added benefit of less sludge production. Single bacterial strains as well as a mixture of different bacterial strains (consortia) are used for this purpose [3, 75]. Coloured and metallic effluents are generally treated with fungal systems on account of the presence of ligninolytic enzymes (laccase, manganese peroxidase & lignin peroxidase) [3, 75]. While Bacillus stearothermophilus strains are used to aerobically degrade azo, Pseudomonas strains are utilized to degrade sulfonated azo dyes [76]. Although these methodologies have many advantages, the optimally promising microbial setting is mandatory for the process to occur [77].

While we have elaborately discussed about the physical, chemical and biological methods, we will now focus on the nanoparticle-based method, which is the topic of recent interest and also the highlight of the current chapter. Semiconductor oxides and plasmonic nanomaterials such as zinc, titanium & iron oxide and silver & gold display effective photocatalytic activity as well as fast oxidation. This aids in the treatment of hazardous dyes in wastewater [78] on account of unique physiochemical properties (such as the great surface area to volume ratio, minor diffusion resistance, fast adsorption equilibrium & capacity) and is hence a quickly growing sector of dye effluent treatment [78, 79]. Nano-silica powder synthesized in combination with silver nanoparticles is reported to aid in 99% elimination of dye with an initial concentration of 50 mg/L [80]. Recently, magnetic nanomaterials have been widely adopted (in synergy with plasmonic nanomaterials) for dye degradation as they can be easily separated by subjecting to a magnetic field [3, 81 - 83]. Dye effluents with carboxylic, hydroxyl, amide functional groups are effectively treated with carbon nanotubes on account of the formation of strong hydrogen bonds [3, 84 - 86]. Recently, graphene oxide and reduced graphene oxide, as well as their hybrid combinations with dielectric and plasmonic nanomaterials, have been exponentially explored for photocatalytic degradation of different dye systems [87 - 90]. It has been observed that over the years, different nanomaterials and nanohybrids are studied in different forms such as catalytic membrane, absorptive, bioactive nanoparticles, biomimetic membrane, polymeric and nano composite membrane as well as thin film composite membrane [3, 91].

In summary, different strategies, including membrane filtration (micro-, nano- and ultra-filtration and reverse osmosis), adsorption, advanced oxidation process, bio-electrochemical treatments, microbial technologies (bacterial, algal, yeast-based and fungal biodegradation) and photocatalytic degradation for reduction of dye-based pollutants from industrial effluent are extensively explored [2 - 4]. However, the nanoparticle and nanohybrid based techniques *via* surfactant free routes are in a nascent stage and demand more experimentation and analysis to gather considerable information about the advantages and disadvantages. Table **6** presents the widely used techniques for dye removal along with their advantages and disadvantages, thereby providing numerous opportunities for future research in this direction.

Table 6. Merits and demerits of the different dye removal methods [2 - 4, 8].

S. No.	Methods Developed	Merits	Demerits
1	Photochemical	No sludge generation	Production of by-products
2	NaOCl	Hastens the azo bond cleavage	Discharge of aromatic amine
3	Ozonation	Applicable in a gas state	Short half-life
4	Cucurbituril	Exhibits good adsorption and removal of various dyes	Expensive
5	Activated carbon	Note worthy removal of a wide variety of dyes	Very expensive
6	Peat	Good adsorbent due to cellular structure	Specific surface areas for adsorption are poorer than that of activated carbon
7	Electrochemical destruction	Breakdown compounds are not hazardous in nature	High cost involved in the electricity
8	Irradiation	Excellent oxidation at the lab scale	Requires a lot of dissolved O_2 for a large-scale study
9	Fentons reagent	Both soluble and insoluble dyes can be treated	Sludge generation
10	Wood chips	Exhibit good adsorption capacity for acid dyes	Necessitate long retention times
11	Membrane filtration	Removes all, more or less, dye types	Generation of concentrated sludge
12	Silica gel	Exhibits good adsorption capacity for basic dyes	Side reactions avoid use in a commercial application
13	Ion exchange	Effective Regeneration of the adsorbent is possible	Not effective for all dye system

(Table 6) cont.....

S. No.	Methods Developed	Merits	Demerits
14	Techniques elaborated in this chapter based on surfactant-free nanohybrids	High surfaces are of adsorbents on account of graphene parent structure	Regeneration of nanohybrids and effect of pH, surfactant, temperature and competing ions/molecules demand further experimental analysis

There are many publications reviewing different strategies that have been developed in the past for general dye degradation. From the extensive literature review, it may be noted that every methodology that has been developed for dye degradation has certain advantages and disadvantages with regard to aspects including: effectiveness of the strategy under various conditions, practicality for industrial water treatment, necessities of pre- and post-treatment, rate of removal and cost involved as well as the impact on the environment. Given these intricate complications, a methodology that can simultaneously satisfy all the above-mentioned conditions is ideally sought-after and remains the most difficult task to achieve. In this context, a combination of existing methods and under-development methods is explored so as to maximize the merits and minimize the demerits. Further, we emphasize that the results documented in this chapter with regard to the use of surfactant-free nanocomposites based on GO and plasmonic NPs are in the embryonic stage and require elaborated experiments with field trials to establish them as an effective strategy for industrial wastewater treatment. Hence, it may be categorized as an under-development strategy for dye degradation. In this context, the study pertaining to % dye removal, the efficiency of the process and possibilities of reusing the absorbent, as well as the effect of various parameters, such as pH, surfactant, temperature and competing ions/molecules, needs to be performed to provide better insights. We believe that the combination of the strategies developed in this chapter with the existing well-established methodologies would aid in better performance of techniques toward dye degradation applications.

CONCLUDING REMARKS

In summary, Au-LCG synthesized by a surfactant free process showed excellent catalytic performance in the reduction of 4 textile and laser dyes. In the case of MB alone, the reaction showed 17000 times better catalytic activity compared to the uncatalyzed reaction. In the case of the reduction of RB, the catalyst has shown 1.7 times better catalytic activity compared to citrate capped commercial counterparts of similar size. This remarkable catalytic activity is attributed to the unmasked, pristine catalyst surface and the large surface area of graphene. The catalyst exhibited molecular recognition for various dyes, as evident from their rates not being similar. These enhanced rates with Au-LCG make it a potential candidate material for wastewater treatment methods at an industrial scale. In the longer run, the LAMS technique can be applied to synthesize unique graphene-based materials that have potentially diverse applications of dye degradation and decolorization reactions of molecules that cause environmental and human health adversities.

Furthermore, this chapter also details about the utility of the hybrid soret nano-assembly of Ag and GO to give unique plasmonic properties. These are synthesized using the adiabatic cooling methodology presenting nanovoids and nanocavities in and around the nano-assembly. These regions sustain localized as well as delocalized plasmons that synergistically interact with the π-plasmons of GO to give advanced functionalities. These materials exhibit excellent dye degradation functionality. Hence, this chapter emphasizes another route of soret nano-assemblies *via* adiabatic cooling to realize hybrid nanomaterials without any additional reducing and capping agents. We believe that this work would be of utility to the broad audience of nanoscience and nanotechnology designing experiments for dye degradation applications. This chapter also presents different futuristic opportunities for researchers to synthesize and explore novel nanomaterials with elements from different parts of the periodic table to achieve desired functionality toward dye degradation applications.

ACKNOWLEDGEMENTS

The authors acknowledge support from Tata Education and Development Trust [TEDT/MUM/HEA/SSSIHL/2017- 2018/0069-RM-db], Prasanthi Trust, Inc., U.S. (22-06-2018), DST-Technology Development Program (IDP/MED/ 19/2016), Life Sciences Research Board (LSRB), and DST-Inspire Fellowship (IF180392), Govt. of India. We especially acknowledge SSSIHL-CRIF for extending the usage of the required instrumentation facility. Guidance from Bhagawan Sri Sathya Sai Baba is gratefully acknowledged.

REFERENCES

[1]   E. Routoula, and S.V. Patwardhan, "Degradation of anthraquinone dyes from effluents: A review focusing on enzymatic dye degradation with industrial potential", *Environ. Sci. Technol.,* vol. 54, no. 2, pp. 647-664, 2020.
[http://dx.doi.org/10.1021/acs.est.9b03737] [PMID: 31913605]

[2]   T. Shindhal, P. Rakholiya, S. Varjani, A. Pandey, H.H. Ngo, W. Guo, H.Y. Ng, and M.J. Taherzadeh, "A critical review on advances in the practices and perspectives for the treatment of dye industry wastewater", *Bioengineered,* vol. 12, no. 1, pp. 70-87, 2021.
[http://dx.doi.org/10.1080/21655979.2020.1863034] [PMID: 33356799]

[3]   M. Mehta, M. Sharma, K. Pathania, P.K. Jena, and I. Bhushan, "Degradation of synthetic dyes using nanoparticles: A mini-review", *Environ. Sci. Pollut. Res. Int.,* vol. 28, no. 36, pp. 49434-49446, 2021.
[http://dx.doi.org/10.1007/s11356-021-15470-5] [PMID: 34350572]

[4]   S.A.G.Z. Morsy, A. Ahmad Tajudin, M.S.M. Ali, F.M. Shariff, and F.M. Shariff, "Current development in decolorization of synthetic dyes by immobilized laccases", *Front. Microbiol.*, vol. 11, p. 572309, 2020.
[http://dx.doi.org/10.3389/fmicb.2020.572309] [PMID: 33101245]

[5]   A. Ajmal, I. Majeed, R.N. Malik, H. Idriss, and M.A. Nadeem, "Principles and mechanisms of photocatalytic dye degradation on TiO_2 based photocatalysts: A comparative overview", *RSC Advances,* vol. 4, no. 70, pp. 37003-37026, 2014.
[http://dx.doi.org/10.1039/C4RA06658H]

[6]   N.Y. Donkadokula, A.K. Kola, I. Naz, and D. Saroj, "A review on advanced physico-chemical and biological textile dye wastewater treatment techniques", *Rev. Environ. Sci. Biotechnol.,* vol. 19, no. 3, pp. 543-560, 2020.
[http://dx.doi.org/10.1007/s11157-020-09543-z]

[7]   B. Yu, N. Reddy, B. Liu, Z. Zhu, W. Wang, and C. Hu, "Sequential assembly of PEDOT/$BiVO_4$/FeOOH onto cotton fabrics for photocatalytic degradation of reactive dyes", *Cellulose,* vol. 28, no. 17, pp. 11051-11066, 2021.
[http://dx.doi.org/10.1007/s10570-021-04192-z]

[8]   C. Lavanya, D. Rajesh, C. Sunil, and S. Sarita, "Degradation of toxic dyes: A review", *Int. J. Curr. Microbiol. Appl. Sci.,* vol. 3, pp. 189-199, 2014.

[9]   I.K. Konstantinou, and T.A. Albanis, "TiO_2-assisted photocatalytic degradation of azo dyes in aqueous solution: kinetic and mechanistic investigations", *Appl. Catal. B,* vol. 49, no. 1, pp. 1-14, 2004.
[http://dx.doi.org/10.1016/j.apcatb.2003.11.010]

[10]   M.S. Anantha, S. Olivera, C. Hu, B.K. Jayanna, N. Reddy, K. Venkatesh, H.B. Muralidhara, and R. Naidu, "Comparison of the photocatalytic, adsorption and electrochemical methods for the removal of cationic dyes from aqueous solutions", *Environmental Technology & Innovation,* vol. 17, p. 100612, 2020.
[http://dx.doi.org/10.1016/j.eti.2020.100612]

[11]   K. Thakur, and B. Kandasubramanian, "Graphene and graphene oxide-based composites for removal of organic pollutants: A review", *J. Chem. Eng. Data,* vol. 64, no. 3, pp. 833-867, 2019.
[http://dx.doi.org/10.1021/acs.jced.8b01057]

[12]   S. Marimuthu, A.J. Antonisamy, S. Malayandi, K. Rajendran, P.C. Tsai, A. Pugazhendhi, and V.K. Ponnusamy, "Silver nanoparticles in dye effluent treatment: A review on synthesis, treatment methods, mechanisms, photocatalytic degradation, toxic effects and mitigation of toxicity", *J. Photochem. Photobiol. B,* vol. 205, p. 111823, 2020.
[http://dx.doi.org/10.1016/j.jphotobiol.2020.111823] [PMID: 32120184]

[13]   D. Gola, A. kriti, N. Bhatt, M. Bajpai, A. Singh, A. Arya, N. Chauhan, S.K. Srivastava, P.K. Tyagi, and Y. Agrawal, "Silver nanoparticles for enhanced dye degradation", *Current Research in Green and Sustainable Chemistry,* vol. 4, p. 100132, 2021.
[http://dx.doi.org/10.1016/j.crgsc.2021.100132]

[14]   S. Bhaskar, R. Patra, N. C. S. S. Kowshik, K. M. Ganesh, and V. Srinivasan, "Nanostructure effect on quenching and dequenching of quantum emitters on surface plasmon-coupled interface: A comparative analysis using gold nanospheres and nanostars", *Phys. E,* vol. 124, p. 114276, 2020.

[15]   A. Rai, S. Bhaskar, N. Reddy, and S.S. Ramamurthy, "Cellphone-Aided Attomolar Zinc Ion detection using silkworm protein-based nanointerface engineering in a plasmon-coupled dequenched emission platform", *ACS Sustain. Chem.& Eng.,* vol. 9, no. 44, pp. 14959-14974, 2021.
[http://dx.doi.org/10.1021/acssuschemeng.1c05437]

[16]   S. Rathnakumar, S. Bhaskar, A. Rai, D.V.V. Saikumar, N.S.V. Kambhampati, V. Sivaramakrishnan, and S.S. Ramamurthy, "Plasmon-coupled silver nanoparticles for mobile phone-based attomolar sensing of mercury ions", *ACS Appl. Nano Mater.,* vol. 4, no. 8, pp. 8066-8080, 2021.
[http://dx.doi.org/10.1021/acsanm.1c01347]

[17]   A. Rai, S. Bhaskar, and S.S. Ramamurthy, "Plasmon-coupled directional emission from soluplus-mediated agau nanoparticles for attomolar sensing using a smartphone", *ACS Appl. Nano Mater.*, vol. 4, no. 6, pp. 5940-5953, 2021.
[http://dx.doi.org/10.1021/acsanm.1c00841]

[18]   S. Bhaskar, A.K. Singh, P. Das, P. Jana, S. Kanvah, S. Bhaktha B N, and S.S. Ramamurthy, "Superior resonant nanocavities engineering on the photonic crystal-coupled emission platform for the detection of femtomolar iodide and zeptomolar cortisol", *ACS Appl. Mater. Interfaces*, vol. 12, no. 30, pp. 34323-34336, 2020.
[http://dx.doi.org/10.1021/acsami.0c07515] [PMID: 32597162]

[19]   S. Bhaskar, N.C.S.S. Kowshik, S.P. Chandran, and S.S. Ramamurthy, "Femtomolar detection of spermidine using au decorated SiO_2 nanohybrid on plasmon-coupled extended cavity nanointerface: A smartphone-based fluorescence dequenching approach", *Langmuir*, vol. 36, no. 11, pp. 2865-2876, 2020.
[http://dx.doi.org/10.1021/acs.langmuir.9b03869] [PMID: 32159962]

[20]   A. Rai, S. Bhaskar, G. Kalathur Mohan, and S.S. Ramamurthy, "Biocompatible gellucire ® inspired bimetallic nanohybrids for augmented fluorescence emission based on graphene oxide interfacial plasmonic architectures", *ECS Trans.*, vol. 107, no. 1, pp. 4527-4535, 2022.
[http://dx.doi.org/10.1149/10701.4527ecst]

[21]   S. Bhaskar, and S.S. Ramamurthy, "Synergistic coupling of titanium carbonitride nanocubes and graphene oxide for 800-fold fluorescence enhancements on smartphone based surface plasmon-coupled emission platform", *Mater. Lett.*, vol. 298, p. 130008, 2021.
[http://dx.doi.org/10.1016/j.matlet.2021.130008]

[22]   R.S. Sai Siddhardha, V. Lakshman Kumar, A. Kaniyoor, V. Sai Muthukumar, S. Ramaprabhu, R. Podila, A.M. Rao, and S.S. Ramamurthy, "Synthesis and characterization of gold graphene composite with dyes as model substrates for decolorization: A surfactant free laser ablation approach", *Spectrochim. Acta A Mol. Biomol. Spectrosc.*, vol. 133, pp. 365-371, 2014.
[http://dx.doi.org/10.1016/j.saa.2014.05.069] [PMID: 24967542]

[23]   M. Moronshing, and C. Subramaniam, "Room temperature, multiphasic detection of explosives, and volatile organic compounds using thermodiffusion driven soret colloids", *ACS Sustain. Chem.& Eng.*, vol. 6, no. 7, pp. 9470-9479, 2018.
[http://dx.doi.org/10.1021/acssuschemeng.8b02050]

[24]   S. Bhaskar, P. Das, V. Srinivasan, S.B.N. Bhaktha, and S.S. Ramamurthy, "Plasmonic-Silver Sorets and Dielectric-Nd_2O_3 nanorods for Ultrasensitive Photonic Crystal-Coupled Emission", *Mater. Res. Bull.*, vol. 145, p. 111558, 2022.
[http://dx.doi.org/10.1016/j.materresbull.2021.111558]

[25]   A. Rai, S. Bhaskar, K.M. Ganesh, and S.S. Ramamurthy, "Engineering of coherent plasmon resonances from silver soret colloids, graphene oxide and Nd_2O_3 nanohybrid architectures studied in mobile phone-based surface plasmon-coupled emission platform", *Mater. Lett.*, vol. 304, p. 130632, 2021.
[http://dx.doi.org/10.1016/j.matlet.2021.130632]

[26]   S. Bhaskar, P. Das, M. Moronshing, A. Rai, C. Subramaniam, S.B.N. Bhaktha, and S.S. Ramamurthy, "Photoplasmonic assembly of dielectric-metal, Nd_2O_3-Gold soret nanointerfaces for dequenching the luminophore emission", *Nanophotonics*, vol. 10, no. 13, pp. 3417-3431, 2021.
[http://dx.doi.org/10.1515/nanoph-2021-0124]

[27]   M. Moronshing, S. Bhaskar, S. Mondal, S.S. Ramamurthy, and C. Subramaniam, "Surface-enhanced Raman scattering platform operating over wide pH range with minimal chemical enhancement effects: Test case of tyrosine", *J. Raman Spectrosc.*, vol. 50, no. 6, pp. 826-836, 2019.
[http://dx.doi.org/10.1002/jrs.5587]

[28]   S. Bhaskar, P. Jha, C. Subramaniam, and S. S. Ramamurthy, "Multifunctional hybrid soret

nanoarchitectures for mobile phone-based picomolar Cu^{2+} ion sensing and dye degradation applications", *Phys. E,* vol. 132, p. 114764, 2021.
[http://dx.doi.org/10.1016/j.physe.2021.114764]

[29]   S. Mondal, and C. Subramaniam, "Xenobiotic contamination of water by plastics and pesticides revealed through real-time, ultrasensitive, and reliable surface-enhanced raman scattering", *ACS Sustain. Chem.& Eng.,* vol. 8, no. 20, pp. 7639-7648, 2020.
[http://dx.doi.org/10.1021/acssuschemeng.0c00902]

[30]   S. Bhaskar, M. Moronshing, V. Srinivasan, P.K. Badiya, C. Subramaniam, and S.S. Ramamurthy, "Silver soret nanoparticles for femtomolar sensing of glutathione in a surface plasmon-coupled emission platform", *ACS Appl. Nano Mater.,* vol. 3, no. 5, pp. 4329-4341, 2020.
[http://dx.doi.org/10.1021/acsanm.0c00470]

[31]   S. Stankovich, D.A. Dikin, G.H.B. Dommett, K.M. Kohlhaas, E.J. Zimney, E.A. Stach, R.D. Piner, S.T. Nguyen, and R.S. Ruoff, "Graphene-based composite materials", *Nature,* vol. 442, no. 7100, pp. 282-286, 2006.
[http://dx.doi.org/10.1038/nature04969] [PMID: 16855586]

[32]   S. Park, and R.S. Ruoff, "Chemical methods for the production of graphenes", *Nat. Nanotechnol.,* vol. 4, no. 4, pp. 217-224, 2009.
[http://dx.doi.org/10.1038/nnano.2009.58] [PMID: 19350030]

[33]   J.T. Choi, D.H. Kim, K.S. Ryu, H. Lee, H.M. Jeong, C.M. Shin, J.H. Kim, and B.K. Kim, "Functionalized graphene sheet/polyurethane nanocomposites: Effect of particle size on physical properties", *Macromol. Res.,* vol. 19, no. 8, pp. 809-814, 2011.
[http://dx.doi.org/10.1007/s13233-011-0801-4]

[34]   S. Bhaskar, A. Rai, G. Kalathur Mohan, and S.S. Ramamurthy, "Mobile phone camera-based detection of surface plasmon-coupled fluorescence from streptavidin magnetic nanoparticles and graphene oxide hybrid nanointerface", *ECS Trans.,* vol. 107, no. 1, pp. 3223-3232, 2022.
[http://dx.doi.org/10.1149/10701.3223ecst]

[35]   C. Huang, C. Li, and G. Shi, "Graphene based catalysts", *Energy Environ. Sci.,* vol. 5, no. 10, pp. 8848-8868, 2012.
[http://dx.doi.org/10.1039/c2ee22238h]

[36]   O.G. Apul, Q. Wang, Y. Zhou, and T. Karanfil, "Adsorption of aromatic organic contaminants by graphene nanosheets: Comparison with carbon nanotubes and activated carbon", *Water Res.,* vol. 47, no. 4, pp. 1648-1654, 2013.
[http://dx.doi.org/10.1016/j.watres.2012.12.031] [PMID: 23313232]

[37]   Z. Sun, Z. Rong, Y. Wang, Y. Xia, W. Du, and Y. Wang, "Selective hydrogenation of cinnamaldehyde over Pt nanoparticles deposited on reduced graphene oxide", *RSC Advances,* vol. 4, no. 4, pp. 1874-1878, 2014.
[http://dx.doi.org/10.1039/C3RA44962A]

[38]   W. Choi, I. Lahiri, R. Seelaboyina, and Y.S. Kang, "Synthesis of graphene and its applications: A review", *Crit. Rev. Solid State Mater. Sci.,* vol. 35, no. 1, pp. 52-71, 2010.
[http://dx.doi.org/10.1080/10408430903505036]

[39]   B.F. Machado, and P. Serp, "Graphene-based materials for catalysis", *Catal. Sci. Technol.,* vol. 2, no. 1, pp. 54-75, 2012.
[http://dx.doi.org/10.1039/C1CY00361E]

[40]   H.C. Schniepp, J.L. Li, M.J. McAllister, H. Sai, M. Herrera-Alonso, D.H. Adamson, R.K. Prud'homme, R. Car, D.A. Saville, and I.A. Aksay, "Functionalized single graphene sheets derived from splitting graphite oxide", *J. Phys. Chem. B,* vol. 110, no. 17, pp. 8535-8539, 2006.
[http://dx.doi.org/10.1021/jp060936f] [PMID: 16640401]

[41]   Y. Shen, and A.C. Lua, "A facile method for the large-scale continuous synthesis of graphene sheets using a novel catalyst", *Sci. Rep.,* vol. 3, no. 1, p. 3037, 2013.

[http://dx.doi.org/10.1038/srep03037] [PMID: 24154539]

[42] A. Kaniyoor, T.T. Baby, T. Arockiadoss, N. Rajalakshmi, and S. Ramaprabhu, "Wrinkled graphenes: a study on the effects of synthesis parameters on exfoliation-reduction of graphite oxide", *J. Phys. Chem. C*, vol. 115, no. 36, pp. 17660-17669, 2011.
[http://dx.doi.org/10.1021/jp204039k]

[43] B. Anand, A. Kaniyoor, S.S.S. Sai, R. Philip, and S. Ramaprabhu, "Enhanced optical limiting in functionalized hydrogen exfoliated graphene and its metal hybrids", *J. Mater. Chem. C Mater. Opt. Electron. Devices,* vol. 1, no. 15, pp. 2773-2780, 2013.
[http://dx.doi.org/10.1039/c3tc00927k]

[44] X. Zhou, X. Huang, X. Qi, S. Wu, C. Xue, F.Y.C. Boey, Q. Yan, P. Chen, and H. Zhang, "In situ synthesis of metal nanoparticles on single-layer graphene oxide and reduced graphene oxide surfaces", *J. Phys. Chem. C,* vol. 113, no. 25, pp. 10842-10846, 2009.
[http://dx.doi.org/10.1021/jp903821n]

[45] C. Wen, M. Shao, S. Zhuo, Z. Lin, and Z. Kang, "Silver/graphene nanocomposite: Thermal decomposition preparation and its catalytic performance", *Mater. Chem. Phys.,* vol. 135, no. 2-3, pp. 780-785, 2012.
[http://dx.doi.org/10.1016/j.matchemphys.2012.05.058]

[46] J. Huang, L. Zhang, B. Chen, N. Ji, F. Chen, Y. Zhang, and Z. Zhang, "Nanocomposites of size-controlled gold nanoparticles and graphene oxide: Formation and applications in SERS and catalysis", *Nanoscale,* vol. 2, no. 12, pp. 2733-2738, 2010.
[http://dx.doi.org/10.1039/c0nr00473a] [PMID: 20936236]

[47] R. Muszynski, B. Seger, and P.V. Kamat, "Decorating graphene sheets with gold nanoparticles", *J. Phys. Chem. C,* vol. 112, no. 14, pp. 5263-5266, 2008.
[http://dx.doi.org/10.1021/jp800977b]

[48] A. Kaniyoor, and S. Ramaprabhu, "Soft functionalization of graphene for enhanced tri-iodide reduction in dye sensitized solar cells", *J. Mater. Chem.,* vol. 22, no. 17, pp. 8377-8384, 2012.
[http://dx.doi.org/10.1039/c2jm16596a]

[49] V. Amendola, and M. Meneghetti, "Laser ablation synthesis in solution and size manipulation of noble metal nanoparticles", *Phys. Chem. Chem. Phys.,* vol. 11, no. 20, pp. 3805-3821, 2009.
[http://dx.doi.org/10.1039/b900654k] [PMID: 19440607]

[50] A. Bulychev, and S. Mobashery, "Class C β-lactamases operate at the diffusion limit for turnover of their preferred cephalosporin substrates", *Antimicrob. Agents Chemother.,* vol. 43, no. 7, pp. 1743-1746, 1999.
[http://dx.doi.org/10.1128/AAC.43.7.1743] [PMID: 10390233]

[51] N. Kozer, Y.Y. Kuttner, G. Haran, and G. Schreiber, "Protein-protein association in polymer solutions: From dilute to semidilute to concentrated", *Biophys. J.,* vol. 92, no. 6, pp. 2139-2149, 2007.
[http://dx.doi.org/10.1529/biophysj.106.097717] [PMID: 17189316]

[52] A. Gangula, R. Podila, R. M, L. Karanam, C. Janardhana, and A.M. Rao, "Catalytic reduction of 4-nitrophenol using biogenic gold and silver nanoparticles derived from Breynia rhamnoides", *Langmuir,* vol. 27, no. 24, pp. 15268-15274, 2011.
[http://dx.doi.org/10.1021/la2034559] [PMID: 22026721]

[53] E.H. Martins Ferreira, M.V.O. Moutinho, F. Stavale, M.M. Lucchese, R.B. Capaz, C.A. Achete, and A. Jorio, "Evolution of the Raman spectra from single-, few-, and many-layer graphene with increasing disorder", *Phys. Rev. B Condens. Matter Mater. Phys.,* vol. 82, no. 12, p. 125429, 2010.
[http://dx.doi.org/10.1103/PhysRevB.82.125429]

[54] S. Moussa, A.R. Siamaki, B.F. Gupton, and M.S. El-Shall, "Pd-partially reduced graphene oxide catalysts (Pd/PRGO): Laser synthesis of Pd nanoparticles supported on PRGO nanosheets for carbon–carbon cross coupling reactions", *ACS Catal.,* vol. 2, no. 1, pp. 145-154, 2012.
[http://dx.doi.org/10.1021/cs200497e]

[55]   S. Moussa, G. Atkinson, M. SamyEl-Shall, A. Shehata, K.M. AbouZeid, and M.B. Mohamed, "Laser assisted photocatalytic reduction of metal ions by graphene oxide", *J. Mater. Chem.,* vol. 21, no. 26, pp. 9608-9619, 2011.
[http://dx.doi.org/10.1039/c1jm11228g]

[56]   S. Moussa, V. Abdelsayed, and M. Samy El-Shall, "Laser synthesis of Pt, Pd, CoO and Pd–CoO nanoparticle catalysts supported on graphene", *Chem. Phys. Lett.,* vol. 510, no. 4-6, pp. 179-184, 2011.
[http://dx.doi.org/10.1016/j.cplett.2011.05.026]

[57]   J.P. Sylvestre, S. Poulin, A.V. Kabashin, E. Sacher, M. Meunier, and J.H.T. Luong, "Surface chemistry of gold nanoparticles produced by laser ablation in aqueous media", *J. Phys. Chem. B,* vol. 108, no. 43, pp. 16864-16869, 2004.
[http://dx.doi.org/10.1021/jp047134+]

[58]   S. Kundu, and H. Liang, "Shape-selective formation and characterization of catalytically active iridium nanoparticles", *J. Colloid Interface Sci.,* vol. 354, no. 2, pp. 597-606, 2011.
[http://dx.doi.org/10.1016/j.jcis.2010.11.032] [PMID: 21144533]

[59]   V.R. Dantham, P.B. Bisht, B.S. Kalanoor, T.T. Baby, and S. Ramaprabhu, "Restricting charge transfer in dye-graphene system", *Chem. Phys. Lett.,* vol. 521, pp. 130-133, 2012.
[http://dx.doi.org/10.1016/j.cplett.2011.11.077]

[60]   Y. Liu, C. Liu, and Y. Liu, "Investigation on fluorescence quenching of dyes by graphite oxide and graphene", *Appl. Surf. Sci.,* vol. 257, no. 13, pp. 5513-5518, 2011.
[http://dx.doi.org/10.1016/j.apsusc.2010.12.136]

[61]   M.M. Hassan, C.J. Hawkyard, and P.A. Barratt, "Decolourisation of dyes and dyehouse effluent in a bubble-column reactor by ozonation in the presence of H_2O_2, $KMnO_4$ or Ferral", *J. Chem. Technol. Biotechnol.,* vol. 81, no. 2, pp. 158-166, 2006.
[http://dx.doi.org/10.1002/jctb.1373]

[62]   K. Wu, Y. Xie, J. Zhao, and H. Hidaka, "Photo-Fenton degradation of a dye under visible light irradiation", *J. Mol. Catal. Chem.,* vol. 144, no. 1, pp. 77-84, 1999.
[http://dx.doi.org/10.1016/S1381-1169(98)00354-9]

[63]   N.R. Wilson, P.A. Pandey, R. Beanland, R.J. Young, I.A. Kinloch, L. Gong, Z. Liu, K. Suenaga, J.P. Rourke, S.J. York, and J. Sloan, "Graphene oxide: structural analysis and application as a highly transparent support for electron microscopy", *ACS Nano,* vol. 3, no. 9, pp. 2547-2556, 2009.
[http://dx.doi.org/10.1021/nn900694t] [PMID: 19689122]

[64]   A. P J, B. Seemesh, R.K.R. G, S.K. P, and R. v, "Disulphide linkage: To get cleaved or not? Bulk and nano copper based SERS of cystine", *Spectrochim. Acta A Mol. Biomol. Spectrosc.,* vol. 196, pp. 229-232, 2018.
[http://dx.doi.org/10.1016/j.saa.2018.02.010] [PMID: 29454250]

[65]   S. Bhaskar, and S.S. Ramamurthy, "Mobile phone-based picomolar detection of tannic acid on Nd_2O_3 nanorod–metal thin-film interfaces", *ACS Appl. Nano Mater.,* vol. 2, no. 7, pp. 4613-4625, 2019.
[http://dx.doi.org/10.1021/acsanm.9b00987]

[66]   A. P J, S. Bhaskar, R.K.R. G, S.K. P, and R. v, "The photocatalytic role of electrodeposited copper on pencil graphite", *Phys. Chem. Chem. Phys.,* vol. 20, no. 5, pp. 3430-3432, 2018.
[http://dx.doi.org/10.1039/C7CP08383A] [PMID: 29340389]

[67]   S. Bhaskar, and S.S. Ramamurthy, "High refractive index dielectric TiO_2 and graphene oxide as salient spacers for >300-fold enhancements", *IEEE International Conference on Nanoelectronics, Nanophotonics, Nanomaterials, Nanobioscience & Nanotechnology (5NANO) IEEE,* pp. 1-6, 2021.
[http://dx.doi.org/10.1109/5NANO51638.2021.9491131]

[68]   A. Rai, S. Bhaskar, K.M. Ganesh, and S.S. Ramamurthy, "Gelucire®-mediated heterometallic AgAu nanohybrid engineering for femtomolar cysteine detection using smartphone-based plasmonics technology", *Mater. Chem. Phys.,* vol. 279, p. 125747, 2022.

[http://dx.doi.org/10.1016/j.matchemphys.2022.125747]

[69]   A. Rai, S. Bhaskar, K.M. Ganesh, and S.S. Ramamurthy, "Cellphone-based attomolar tyrosine sensing based on Kollidon-mediated bimetallic nanorod in plasmon-coupled directional and polarized emission architecture", *Mater. Chem. Phys.,* vol. 285, p. 126129, 2022.
        [http://dx.doi.org/10.1016/j.matchemphys.2022.126129]

[70]   A. Rai, S. Bhaskar, P. Battampara, N. Reddy, and S. Sathish Ramamurthy, "Integrated Photo-Plasmonic coupling of bioinspired Sharp-Edged silver Nano-particles with Nano-films in extended cavity functional interface for Cellphone-aided femtomolar sensing", *Mater. Lett.,* vol. 316, no. June, p. 132025, 2022.
        [http://dx.doi.org/10.1016/j.matlet.2022.132025]

[71]   N.C.L. Beluci, G.A.P. Mateus, C.S. Miyashiro, N.C. Homem, R.G. Gomes, M.R. Fagundes-Klen, R. Bergamasco, and A.M.S. Vieira, "Hybrid treatment of coagulation/flocculation process followed by ultrafiltration in TIO_2-modified membranes to improve the removal of reactive black 5 dye", *Sci. Total Environ.,* vol. 664, pp. 222-229, 2019.
        [http://dx.doi.org/10.1016/j.scitotenv.2019.01.199] [PMID: 30743115]

[72]   S. Sadhasivam, S. Savitha, K. Swaminathan, and F.H. Lin, "Metabolically inactive Trichoderma harzianum mediated adsorption of synthetic dyes: Equilibrium and kinetic studies", *J. Taiwan Inst. Chem. Eng.,* vol. 40, no. 4, pp. 394-402, 2009.
        [http://dx.doi.org/10.1016/j.jtice.2009.01.002]

[73]   R.G. Saratale, G.D. Saratale, J.S. Chang, and S.P. Govindwar, "Bacterial decolorization and degradation of azo dyes: A review", *J. Taiwan Inst. Chem. Eng.,* vol. 42, no. 1, pp. 138-157, 2011.
        [http://dx.doi.org/10.1016/j.jtice.2010.06.006]

[74]   M. Ghaedi, S. Heidarpour, S. Nasiri Kokhdan, R. Sahraie, A. Daneshfar, and B. Brazesh, "Comparison of silver and palladium nanoparticles loaded on activated carbon for efficient removal of Methylene blue: Kinetic and isotherm study of removal process", *Powder Technol.,* vol. 228, pp. 18-25, 2012.
        [http://dx.doi.org/10.1016/j.powtec.2012.04.030]

[75]   R.L. Singh, P.K. Singh, and R.P. Singh, "Enzymatic decolorization and degradation of azo dyes – A review", *Int. Biodeterior. Biodegradation,* vol. 104, pp. 21-31, 2015.
        [http://dx.doi.org/10.1016/j.ibiod.2015.04.027]

[76]   R.C. Senan, and T.E. Abraham, "Bioremediation of textile azo dyes by aerobic bacterial consortium", *Biodegradation,* vol. 15, no. 4, pp. 275-280, 2004.
        [http://dx.doi.org/10.1023/B:BIOD.0000043000.18427.0a] [PMID: 15473556]

[77]   G. Crini, and E. Lichtfouse, "Advantages and disadvantages of techniques used for wastewater treatment", *Environ. Chem. Lett.,* vol. 17, no. 1, pp. 145-155, 2019.
        [http://dx.doi.org/10.1007/s10311-018-0785-9]

[78]   S. Agnihotri, D. Sillu, G. Sharma, and R.K. Arya, "Photocatalytic and antibacterial potential of silver nanoparticles derived from pineapple waste: process optimization and modeling kinetics for dye removal", *Appl. Nanosci.,* vol. 8, no. 8, pp. 2077-2092, 2018.
        [http://dx.doi.org/10.1007/s13204-018-0883-9]

[79]   Y. Ma, Y.M. Zheng, and J.P. Chen, "A zirconium based nanoparticle for significantly enhanced adsorption of arsenate: Synthesis, characterization and performance", *J. Colloid Interface Sci.,* vol. 354, no. 2, pp. 785-792, 2011.
        [http://dx.doi.org/10.1016/j.jcis.2010.10.041] [PMID: 21093869]

[80]   S.K. Das, M.M.R. Khan, T. Parandhaman, F. Laffir, A.K. Guha, G. Sekaran, and A.B. Mandal, "Nano-silica fabricated with silver nanoparticles: Antifouling adsorbent for efficient dye removal, effective water disinfection and biofouling control", *Nanoscale,* vol. 5, no. 12, pp. 5549-5560, 2013.
        [http://dx.doi.org/10.1039/c3nr00856h] [PMID: 23680871]

[81]   L. Kong, X. Gan, A.L. Ahmad, B.H. Hamed, E.R. Evarts, B.S. Ooi, and J.K. Lim, "Design and synthesis of magnetic nanoparticles augmented microcapsule with catalytic and magnetic

bifunctionalities for dye removal", *Chem. Eng. J.,* vol. 197, pp. 350-358, 2012.
[http://dx.doi.org/10.1016/j.cej.2012.05.019]

[82] H. Xu, Y. Zhang, Q. Jiang, N. Reddy, and Y. Yang, "Biodegradable hollow zein nanoparticles for removal of reactive dyes from wastewater", *J. Environ. Manage.,* vol. 125, pp. 33-40, 2013.
[http://dx.doi.org/10.1016/j.jenvman.2013.03.050] [PMID: 23643969]

[83] B. Ramalingam, M.M.R. Khan, B. Mondal, A.B. Mandal, and S.K. Das, "Facile synthesis of silver nanoparticles decorated magnetic-chitosan microsphere for efficient removal of dyes and microbial contaminants", *ACS Sustain. Chem.& Eng.,* vol. 3, no. 9, pp. 2291-2302, 2015.
[http://dx.doi.org/10.1021/acssuschemeng.5b00577]

[84] B. Pan, and B. Xing, "Adsorption mechanisms of organic chemicals on carbon nanotubes", *Environ. Sci. Technol.,* vol. 42, no. 24, pp. 9005-9013, 2008.
[http://dx.doi.org/10.1021/es801777n] [PMID: 19174865]

[85] G. Rao, C. Lu, and F. Su, "Sorption of divalent metal ions from aqueous solution by carbon nanotubes: A review", *Separ. Purif. Tech.,* vol. 58, no. 1, pp. 224-231, 2007.
[http://dx.doi.org/10.1016/j.seppur.2006.12.006]

[86] K. Yang, W. Wu, Q. Jing, and L. Zhu, "Aqueous adsorption of aniline, phenol, and their substitutes by multi-walled carbon nanotubes", *Environ. Sci. Technol.,* vol. 42, no. 21, pp. 7931-7936, 2008.
[http://dx.doi.org/10.1021/es801463v] [PMID: 19031883]

[87] M. Al Kausor, and D. Chakrabortty, "Graphene oxide based semiconductor photocatalysts for degradation of organic dye in waste water: A review on fabrication, performance enhancement and challenges", *Inorg. Chem. Commun.,* vol. 129, p. 108630, 2021.
[http://dx.doi.org/10.1016/j.inoche.2021.108630]

[88] F. Khan, M.S. Khan, S. Kamal, M. Arshad, S.I. Ahmad, and S.A.A. Nami, "Recent advances in graphene oxide and reduced graphene oxide based nanocomposites for the photodegradation of dyes", *J. Mater. Chem. C Mater. Opt. Electron. Devices,* vol. 8, no. 45, pp. 15940-15955, 2020.
[http://dx.doi.org/10.1039/D0TC03684F]

[89] H.A.A. Jamjoum, K. Umar, R. Adnan, M.R. Razali, and M.N. Mohamad Ibrahim, "Synthesis, characterization, and photocatalytic activities of graphene oxide/metal oxides nanocomposites: A review", *Front Chem.,* vol. 9, p. 752276, 2021.
[http://dx.doi.org/10.3389/fchem.2021.752276] [PMID: 34621725]

[90] P.B. Sreelekshmi, R.R. Pillai, and A.P. Meera, "Controlled synthesis of novel graphene oxide nanoparticles for the photodegradation of organic dyes", *Top. Catal.,* vol. 65, no. 19-20, pp. 1659-1668, 2022.
[http://dx.doi.org/10.1007/s11244-022-01600-x]

[91] H. Samanta, R. Das, and C. Bhattachajee, "Influence of nanoparticles for wastewater treatment-a short review", *Austin Chem Eng,* vol. 3, p. 1036, 2016.

CHAPTER 4

Treatment of Textile Dye Effluent by Electrochemical Method

Venkatesan Sethuraman[1], Karupannan Aravindh[2], Perumalsamy Ramasamy[2], Bosco Christin Maria Arputham Ashwin[3] and Paulpandian Muthu Mareeswaran[4,*]

[1] *Research and Development, New Energy Storage Technology, Lithium-Ion Battery Division, Amara Raja Batteries Ltd, Karakambadi, Andhra Pradesh-517 520, India*

[2] *Research Centre, SSN College of Engineering, Kalavakkam, Chennai-603110, Tamil Nadu, India*

[3] *Department of Chemistry, Pioneer Kumarasamy College, Nagarcoil-629003, Tamil Nadu, India*

[4] *Department of Chemistry, College of Engineering, Anna University, Chennai-600025, Tamil Nadu, India*

Abstract: This chapter discusses the electrochemical aqueous solution-based breakdown of synthetic textile colours. Several dyeing and finishing industries produce a significant amount of dye wastewater. For the treatment of effluent water, the electrochemical technique is being studied. The discharge of textile wastewater likewise rises as there are more textile industries. So, in recent years, the electrochemical degradation of industrial effluents has gained popularity. Conductivity, pH, process detention times, total suspended solids (TSS), heavy metals, emulsified oils, bacteria, and other pollutants from water are operating factors in electrochemical treatment. Utilizing cyclic voltammetry (CV), reactive synthetic textile dyes' electrochemical behaviour has been reviewed. Studies on chemical oxygen demand (COD), UV-Vis, and CV are chosen to assess the effectiveness of degradation. There are numerous additional businesses that require electrochemical technologies for purifying effluent water. Metal recovery, tanneries, electroplating, dairies, textile processing, oil and oil in water emulsion, and other businesses are among them.

Keywords: Cyclic voltammetry, Chemical oxygen demand, Dyes, Electrochemical degradation, Effluents.

INTRODUCTION

The accumulation of persistent organic pollutants (POPs) in water has grown to be a significant environmental issue during the past two decades, due to modernisation [1 - 3]. Even in smaller quantities (*i.e.*, nanogram to milligram), the

---

* **Corresponding author Paulpandian Muthu Mareeswaran:** Department of Chemistry, College of Engineering, Anna University, Chennai-600025, Tamil Nadu, India; E-mail: muthumareeswaran@gmail.com

Paulpandian Muthu Mareeswaran & Jegathalaprathaban Rajesh (Eds.)

presence of these compounds creates undesirable effects including toxicity, carcinogenicity, mutagenicity, *etc.* [4 - 6]. Even after treatment, these compounds continue to produce toxicity because they are resistant to traditional wastewater treatment techniques [7]. In order to recover and utilize significant water resources, new and effective POP removal technologies are needed. Electrochemical advanced oxidation processes (EAOPs) have been highlighted as one of the potential subcategories of alternate water treatment technologies to address this environmental problem. Because they could handle extremely refractory organic pollutants including drugs [8], insecticides [9], azo dyes [10], and even carboxylic acids [11], EAOPs have gained growing attention. In addition to effectively removing POPs, the EAOPs also have a number of environmental benefits, including (i) mild operating conditions at ambient temperature and pressure; (ii) compact reactors with smaller physical footprints that require less land space; (iii) no additional need for auxiliary chemicals, such as transportation and storage; (iv) do not produce secondary waste streams that need further treatment; and (v) could be easily combined [12]. With all these crucial features, EAOPs are low-carbon, ecologically beneficial technology. Due to its adaptability and simplicity in scaling, anodic oxidation (AO) or electrochemical oxidation (EO) is the EAOP that has received the most attention [13]. Only a few studies have been focused on synthetic water matrices and actual wastewater effluents, while the majority of experimental investigations on EO and other EAOPs in the literature deal with the oxidation of POPs in synthetic wastewater. In order to remediate synthetic water matrices and actual wastewater effluents at the laboratory and pilot plant sizes, the current book chapter tries to explore the possible applicability and limitations of EO methods. For a better grasp of how EO technology works and its potential ramifications, its fundamentals are briefly reviewed. The foundation for POP remediation in the environment is also examined, as is the possibility of combining it with other pre- or post-treatment water technologies.

Depending on the electrode material or electrocatalyst utilized in electrooxidation (EO), electrogenerated ROS [Eq. (1,2)], either physiosorbed or chemisorbed •OH, oxidize organic molecules in the anode area. As a result, some anode materials interact strongly with the produced radical, encouraging oxidation to superoxide (M=O) or chemisorbed oxygen [14]. As "active anodes," these substances (such as platinum, mixed metal oxides based on Ru and Ir, and graphite-carbon electrodes) have low oxygen evolution overpotentials and can only achieve a soft degradation of organic pollutants (electrochemical conversion) with a very low mineralization degree [15]. Other anodes (so-called "nonactive"), such as doped PbO_2, SnO_2, boron-doped diamond, and sub-stoichiometric titanium oxide (Ti_xO_{2x-1}), have a higher oxygen evolution overpotential and weakly interact with the produced radicals, allowing them to freely react with organic molecules until mineralization (electrochemical combustion) [16]. During anodic oxidation processes, a large number of extra oxidants, including ozone, hydrogen peroxide, and peroxosalts, are produced on the anode surface. They interact with organics in large quantities and are effectively in charge of these activities.

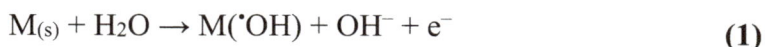

$$M_{(s)} + H_2O \rightarrow M(^{\cdot}OH) + OH^- + e^- \qquad (1)$$

$$M(^{\cdot}OH) \rightarrow M{=}O + H^+ + e^- \qquad (2)$$

Reactive chlorine species (RCS) (Cl_2, HClO, and/or ClO, which are predominant at pH 3.0, 3.0 - 6.5, and > 8.0, respectively) Fig. (**1**) are produced in bulk solution *via* reactions (Eq. (3-5)) when treating solutions containing high concentrations of Cl ions (*i.e.*, reverse osmosis concentrates) along with reactive oxygen species (ROS) [17]. The RCSs are moderately powerful oxidants that can break down various kinds of organic pollutants. However, this approach usually results in the formation of dangerous by-products such haloacetic acids and trihalomethanes as well as other refractory organochlorinated intermediates that are challenging to mineralize [18].

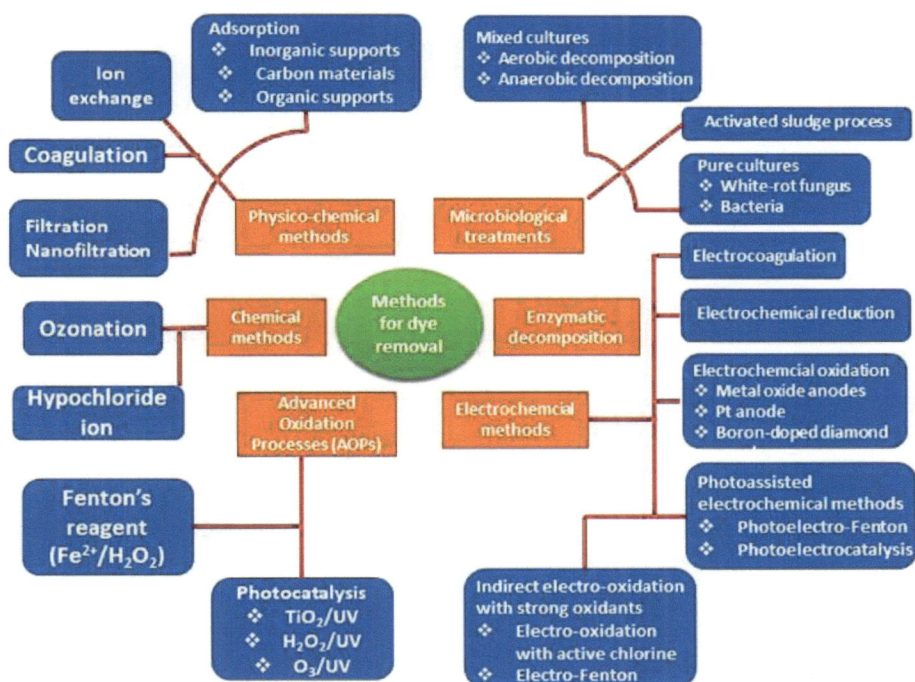

Fig. (1). Main methods used for the removal of organic dyes from wastewaters.

$$2\ Cl^- \rightarrow Cl_{2(aq)} + 2\ e^- \qquad (3)$$

$$Cl_{2(aq)} + H_2O \rightarrow HClO + Cl^- + H^+ \qquad (4)$$

$$HClO \leftrightarrow ClO^- + H^+ \qquad (5)$$

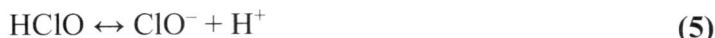

Traditional physico-chemical techniques for cleaning dyeing wastewaters include adsorption with inorganic (typically activated carbon materials) and organic supports, coagulation with lime, aluminum, or iron salts, filtration, and ion exchange (Fig. **1**). These procedures effectively remove colour, but their application is constrained by the need to regularly regenerate the adsorbent materials or the formation of waste sludge that must be disposed of [19]. Fast decolorization and dye degradation are achieved by AOPs like Fenton's reagent and photocatalytic systems involving TiO_2/UV, H_2O_2/UV, and O_3/UV, as well as more potent chemical techniques like ozonation and hypochlorite ion oxidation, as shown in Fig. (**1**) [20]. These methods are not currently in widespread usage, nevertheless, due to their high cost and operational problems. On the other hand, the appealing and simple method of using microorganisms in the biodegradation of synthetic colours. A variety of activated sludge techniques, mixed cultures with aerobic and anaerobic decomposition, and pure cultures with white-rot fungus and bacteria have all been used to investigate decolorization and dye degradation. Fig. (**1**) also depicts the primary electrochemical techniques for dyestuff wastewater treatment. Electrocoagulation (EC), electrochemical oxidation (EO) with different anodes, and indirect electro-oxidation with active chlorine are the typical methods used to eliminate these contaminants.

The potential function of electrochemical reduction has received much less attention recently than new technologies like electro-Fenton (EF) and photo-assisted systems like photoelectro-Fenton (PEF) and photoelectrocatalysis. Electrochemical oxidation, electro-Fenton, and photoassisted electrochemical systems (EAOPs) are examples of electrochemical advanced oxidation processes. In order to better comprehend the advantages and disadvantages of technology in terms of mitigating environmental contamination from synthetic organic dyes, the basics of technology are also briefly reviewed.

Fig. (**2**) depicts this classification for numerous common dyes, along with their chemical make-up and/or colour-index names. The latter nomenclature is more frequently used, and it consists of the name of a general attribute that characterizes the product, followed by the name of its colour and an order number. Acidic (negatively charged), Basic (positively charged), Reactive (the anionic dye used in the textile industry), Mordant (a metallic ion is required for showing their colour or staining selectivity), Vat (it derives from natural indigo),

Disperse (non-ionic dye used in aqueous dispersion), and so on are examples of first names.

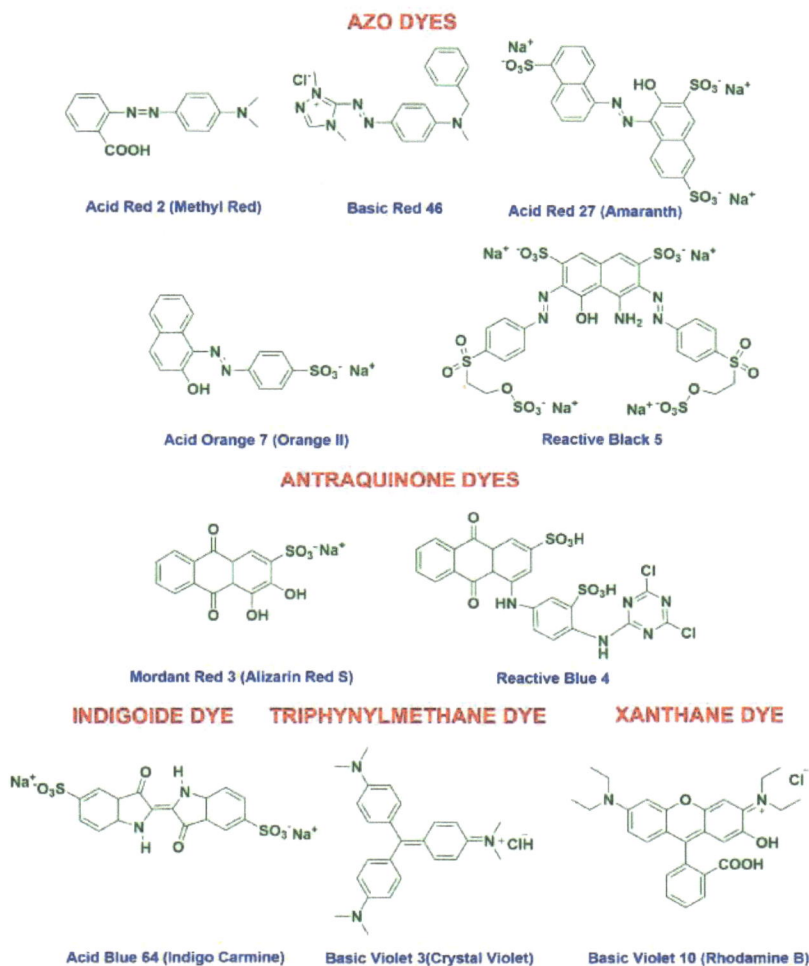

Fig. (2). Chemical structure of typical synthetic organic dyes classified by their chromophore group. The color index and/or common (between parentheses) name of each dye.

Electrocoagulation

The traditional physico-chemical phase separation method of coagulation is used to clean dye wastewater before they are released into the environment. Coagulating substances like Fe^{3+} or Al^{3+} ions, typically in the form of chlorides, are added to aid in the precipitation of dyes. Using electrochemical technology, the EC strategy can produce comparable results [21]. This method, which is depicted in Fig. (**3**), uses a current to dissolve Fe (or steel) or Al sacrificial anodes that are submerged in contaminated water. This process produces metal ions that give separate Fe(II) (and/or Fe(III)) or Al(III) species with hydroxide ions depending on the pH of the medium [22].

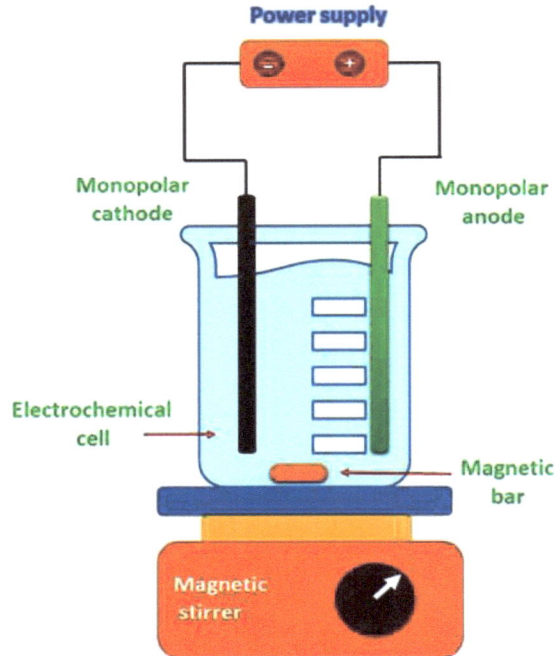

Fig. (3). Electrochemical cells with stirred tank reactors for treating dyes by electrocoagulation in batch operation mode.

These species function as destabilizers or coagulants, neutralizing charges and enable the separation of colours from wastewater. The coagulated particles may be separated by electroflotation when they are linked to the H_2 gas bubbles created at the cathode and driven to the top of the solution. In general, the following significant events take place during an EC treatment [23]: (i) electrode reactions to produce metal ions from Fe or Al anodes and H_2 gas at the cathode; (ii) formation of coagulants in the wastewater; (iii) removal of dyes with coagulants by sedimentation or by electroflotation with evolved H_2; (iv) other electrochemical and chemical reactions involving the reduction of organic impurities and metal ions at the cathode and coagulation of colloidal particles.

Many advantages for EC have been reported [24]: (i) more effective and rapid organic matter separation than in coagulation; (ii) pH control is not necessary, except for extreme values; (iii) the amount of chemicals required is small; (iv) the amount of sludge produced is smaller when compared with coagulation. For example, the sludge formed in the EC method with Fe contains a higher content of dry and hydrophobic solids than that produced in coagulation by the action of $FeCl_3$ followed by the addition of NaOH or lime; (v) the operating costs are much lower than in most conventional technologies.

However, this method presents as major disadvantages [25]: (i) anode passivation and sludge deposition on the electrodes that can inhibit the electrolytic process in continuous operation mode; (ii) high concentrations of iron and aluminium ions in the effluent that must be removed. In the subsections below, the characteristics of EC and its relevant applications to dye removal will be discussed based on the sacrificial anode used.

Coagulation of these flocs produces particles that are removed from the wastewater by sedimentation or electroflotation. The dye can act as a ligand to bind a hydrous iron moiety of the floc, generating a surface complex. From these foundational concepts, a variety of devices and reactors have been developed to implement this process, some of which are even functional for industrial wastewater [26]. There are several bench-scaled stirred tank reactors and filter-press flow systems, respectively, operating in batch or continuous mode. The electrodes utilized are parallel plate electrodes with a monopolar or bipolar connection. Monopolar electrodes have the same polarity on both faces and require an external electrical connection to the power source. On the other hand, bipolar electrodes are positioned between two end monopolar electrodes and lack any electrical connections. The intermediate bipolar electrodes polarize as a function of the power supply's voltage between the subsequent electrodes, producing different polarities on the opposite sides [27, 28].

Experimentally, the decolorization efficiency or percentage of color removal during the treatment of dyes wastewaters is determined by the expression:

$$Color\ Removal\ (\%) = \frac{ABS_0^M - ABS_0^M}{ABS_0^M} \times 100 \qquad (6)$$

where ABS^M_0 and ABS^M_t are the average absorbances before electrolysis and after an electrolysis time t, respectively, at the maximum visible wavelength (l_{max}) of the wastewater. The decontamination process of the dye wastewater is monitored from the abatement of its chemical oxygen demand (COD) and/or total organic carbon (TOC). From these data, the percentages of COD and TOC decays are calculated from the following equations.

$$COD\ decay\ (\%) = \frac{\Delta COD}{COD_0} \times 100 \qquad (7)$$

$$TOC\ decay\ (\%) = \frac{\Delta TOC}{TOC_0} \times 100 \qquad\qquad (8)$$

Electrochemical Reduction

There have been a few articles that discuss the direct electroreduction of dyes in aqueous solution on suitable cathodes. The lack of interest in this conventional electrochemical method is due to the fact that, when compared to more potent direct and indirect electro-oxidation approaches, it offers insufficient wastewater purification. The electrochemical treatment of the azo dye is carried out at room temperature using the standard three-electrode two-compartment cell [29]. A magnetic bar stirrer is used to stir the contaminated solution continuously in the working compartment, and a Pt wire and a by-product from the electroreduction of the chromophoric -N N - group to produce an NH-NH bond are used in the counter electrode compartment. The same research team earlier described a similar behaviour for the electroreduction of the azo dye Reactor fix Golden Yellow 3 under the same experimental conditions [30]. A salt bridge connects the working compartment to the saturated calomel electrode (SCE), which serves as the reference electrode. The electroreduction of 80 mg dm^3 Amaranth solutions in 0.1 M Na$_2$SO$_4$ with a phosphate buffer of pH 6.6 on ACF was contrasted using this technology under galvanostatic and potentiostatic conditions [31, 32]. After 500 min of potentiostatic electrolysis, Fig. (4) shows the connection between colour, COD, and TOC reductions with applied electrode potential [32]. With a cathodic potential (E$_{cat}$) > 0.500 V vs. SCE, electrochemical reduction results in around 60% reductions in COD and TOC, as well as general decolorization. Analysis of the spectrum for E$_{cat}$ showed that the azo bond was destroyed between 0.200 and 0.800 V. Operating at a constant j of 1.0 mA cm2 for 8 hours produced results that were comparable, including 95% colour removal and 62% COD decay. The insufficient COD and TOC reductions for Amaranth solutions were attributed to the adsorption of byproducts on ACF during the electroreduction process [31, 32].

According to a study by Nagaraj *et al.* [33], significant treatment of the majority of contaminants is required in order to make the water reusable, necessitating the employment of multiple challenging procedures. In the current study (GCE), graphene oxide (GO) nanoparticles made on a glassy carbon electrode were used to electrochemically breakdown the azo dye Congo red (CR). Since the new method can be used to examine CR dye in both soil and water samples, it has drawn a lot of interest. Voltammetric techniques were employed to analyze dye electrochemical performance, and a linear relationship in the 0.01 to 0.2 M range with an LOD value of 2.4×10^{-7} M was found. The expanded CR inquiry can benefit from the enhanced sensor's lower detection limit value.

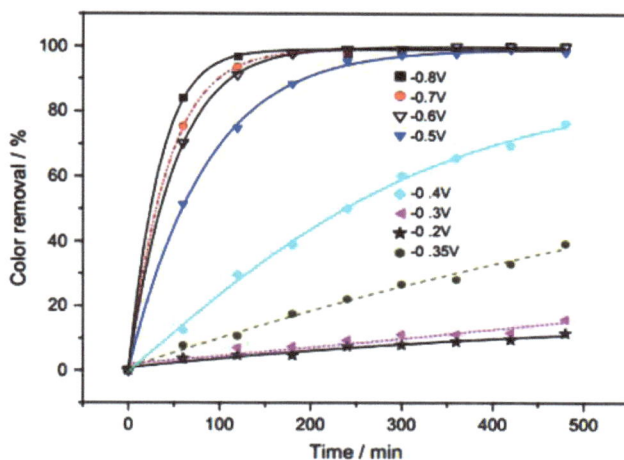

Fig. (4). Dependence of color removal on time for amarnath solution on ACF with electroreduction.

Zhou *et al.* [34] found low decolorization efficiency and TOC removal. The practical application electro-Fenton (EF) process is severely constrained by the performance of the cathode in H_2O_2 electrogeneration. Here, we describe an electrochemical modification of reticulated vitreous carbon foam (RVC foam) electrode for improved H_2O_2 electrogeneration that is both straightforward and efficient. The redesigned electrode was examined using cyclic voltammetry, chronoamperometry, and X-ray photoelectron spectroscopy. When oxygen-containing groups (72.5-184.0 mol/g) were added to the surface of RVC foam, the yield of H_2O_2 increased by 59.8-258.2%. The redesigned electrodes demonstrated excellent stability and significantly increased electrocatalytic activity for O_2 reduction. Additionally, a pulsed current method was suggested with the intention of reducing the degree of electroreduction of H_2O_2 in porous RVC foam. In comparison to the original and changed electrodes, the H_2O_2 concentration was 582.3 and 114.0% higher, respectively. Reactive Blue 19 (RB19) was removed using the EF technique to verify the viability of alteration and pulsed current (Fig. **5**). More hydroxyl radicals were produced, as evidenced by the fluorescence intensity of hydroxybenzoic acid being 3.2 times higher in EF with the changed electrode than in EF with the unmodified electrode. The removal efficiencies of RB 19 in EF with a non-modified electrode, a modified electrode, and a modified electrode assisted by pulsed current were 53.9, 68.9, and 81.1%, respectively. This data shows that pulsed current and the green modification approach are both applicable for pollutant removal in EF systems.

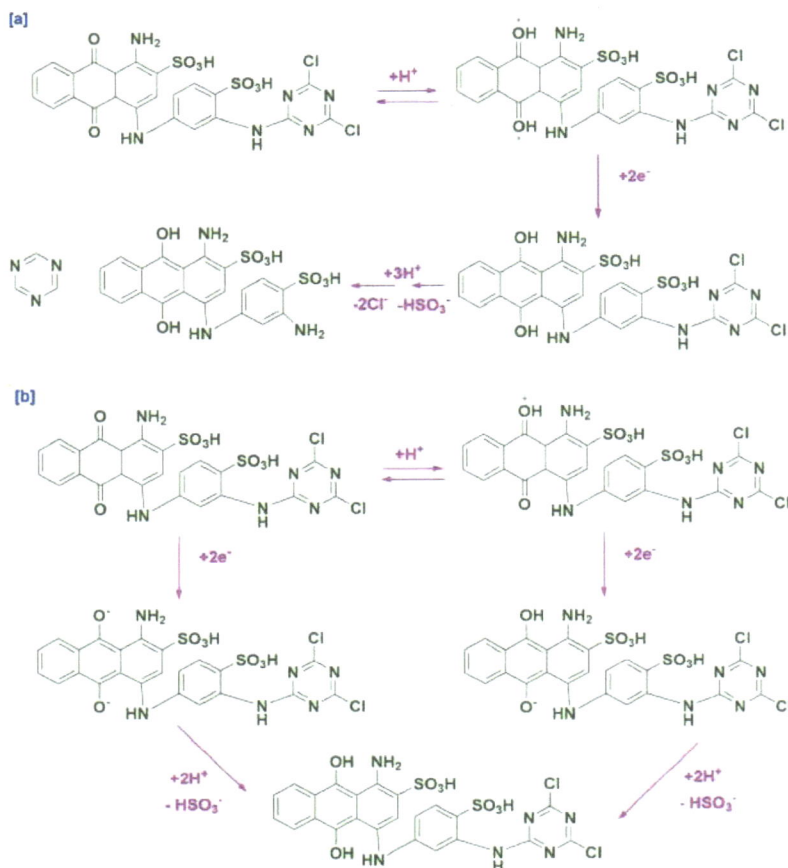

Fig. (5). Electrochemical reduction of Reactive Blue 4 on reticulated glassy carbon electrode at (a) pH < 8.0 and (b) pH > 8.0. Adapted from Ref. [35].

Metal Anodes

Rapid decolorization and extensive purification of extremely concentrated dye effluents can also be achieved by using Ti/Pt in indirect electro-oxidation with active chlorine. Following their comparative investigation of the breakdown of reactive orange 4 in Na_2SO_4 and NaCl effluents, Lopez-Grimau and Gutierrez [36] have demonstrated this. To treat 4 dm^3 of 1 g/dm^3 of this dye in 20 g/dm^3 of NaCl at pH 9.0, 24 mA/cm^2, and a flow rate of 25 dm^3h^{-1} in batch mode, a flow cell with a 486 cm^2 Ti/Pt anode flow rate was used. The total colour of the effluents was removed in 90 minutes, but mineralization took significantly longer. After 10 hours of treatment, 81% of the COD and TOC had decayed, at a very high energy cost of 698 kWh/m^3. These observations led the scientists to the conclusion that since conventional biological treatment is unable to successfully remove the colour of dyeing effluents, this approach can only be appealing for decolorizing those substances. In a similar manner, Vlyssides *et al.* [37] recommended using facilitated electro-oxidation with a Ti/ Pt anode as a pre-treatment stage for biological post-treatment of textile dyestuff wastewaters with high Cl concentration.

The performance of stirred undivided [38] and divided [39] tank reactors with a 100 cm^2 Ti plate coated with Pt (70 mol%) and Ir (30 mol%) as a metal anode and a 100 cm^2 stainless cathode has recently been tested for the decolorization of 600-800 cm^3 of 0.6 g dm3 of the reactive dye Red Procion H-EXGL in 16.25 g/dm^3 NaCl at pH 11. These tests showed that the dye only reacts with active chlorine agents in the bulk of the solution and is not oxidized at the Ti/Pt-Ir anode. Due to the acceleration of all electrode and chemical interactions in the undivided cell, the entire decolorization was achieved more quickly with a temperature increase from 10 to 40 °C without a substantial pH change. When the pH of the anolyte of the divided cell unexpectedly dropped to around 6, uncompensated reactions like the direct oxidation of chloride ions at the anode to produce soluble chlorine and favouring the dye oxidation with HClO instead of its slower reaction with ClO taking place in the alkaline medium led to a much faster colour removal being discovered. The rate of dye decay in this system increased up to 40 C when the temperature was raised from 10 C to 80 C; after that point, it gradually slowed down due to a decrease in the solubility of Cl$_2$(aq). Up to 40 °C, the divided cell required less energy to remove all of the colour, whereas higher temperatures required less energy for the undivided system. These results demonstrate the necessity of changing the structure of the electrolytic system as well as variables like pH and temperature in order to maximize the oxidation ability of electrogenerated active chlorine.

Graphite Anode

Numerous organic dyes, including Indigo [40], Congo Red [41], and Methyl Orange [42], have been decoloured *via* indirect electro-oxidation with active chlorine and a graphite anode. The two latter, for instance, *demonstrated* rapid colour removal that increased with increasing Cl concentration, current density, temperature (from 20–25 to 40–45 C), and pH (as expected if they react with Cl$_2$(aq), HClO, and/or ClO with only a minimal contribution from their direct anodic oxidation on graphite). A scale-up experiment conducted by Cameselle *et al.* [40] for the treatment of 0.2 g/dm^3 indigo in 30 g/dm^3 NaCl at natural pH utilizing 20 dm^3 solutions in a stirred undivided graphite/graphite cell with 500 cm^2 electrodes separated by 50 cm was also apparent. 90% decolorization efficiency was found after using a cell voltage of 5 V at room temperature for 2 hours, with a very low energy usage of 1.86 kWh m^3. This supports once more the viability of utilizing electrogenerated active chlorine in mediated oxidation as a highly appealing technique for the effective and affordable colour removal of industrial dyeing wastewater.

In contrast, DSA-type cathodes, Pt, and carbonaceous materials wastefully remove shorter aromatics and aliphatic acids from arrangements whereas non-active PbO_2 and conducting jewel cathodes can act as though they are mineralizing shorter aromatics and aliphatic acids formed throughout the procedure. Among the two non-active anodes, jewel terminals seem most suited for use in honing due to their greater ability to completely mineralize colour combinations with the least amount of energy consumption. However, more research is needed to fully understand a few crucial points, such as the benefit lifetime of large-scale BDD electrodes, an addition to financial analysis of the method for real wastewaters, the optimization of electrolytic reactors, *etc.* The reasonability of EO with these electrodes for mechanical application is not well established. The production of more of the most potent heterogeneous oxidant, OH, and other ROS from water discharge, which might quickly react with organics as much as general mineralization, is what gives diamond anodes their highest oxidation electricity. Additional technologies of weaker peroxo derivative oxidants made from the electrolytes bisulphate, bicarbonate, and phosphate at those anodes can be crucial to the combustion process in chloride-free solutions. While chloride is present in treated wastewaters, active chlorine species are effortlessly created in all anodes, accelerating the breakdown of dyes and, in most instances, significantly enhancing the electrolytic device's capacity for oxidation. Comparative studies using the extraordinary electrolytes recommended in this segment have demonstrated that many dyestuffs are destroyed quickly in chloride-containing effluents.

Boron-doped Diamond Electrodes

Due to their numerous technologically significant properties, such as an inert surface with low adsorption properties, remarkable corrosion stability even in strongly acidic media, and extremely high O_2 evolution overvoltage, BDD thin films—relatively new electrode materials—have drawn a lot of attention [43]. These qualities make them good materials for EO. Since it is anticipated that BDD anodes will not offer any catalytically active sites for the adsorption of reactants and/or products in aqueous medium, they have been classified as non-active electrodes. The electrochemical combustion of organic pollutants is then attributed to the hydroxyl radical (BDD(OH)) formed from water discharge on their surface reaction, though slower reactions with other ROS (H_2O_2 and O_3) and weaker electrogenerated oxidants (peroxodisulphate, peroxodicarbonate, or peroxodiphosphate) are also possible.

Numerous studies have shown that using a BDD thin film in EO allows for total mineralization of various organics in real wastewaters with good current efficiency [44]. Recently, this synthetic material was used to handle dyestuff and was deposited on various supports. It is significant to note that the majority of investigations have used Si-supported devices [45], despite the challenges associated with their industrial application due to the brittleness and relatively low conductivity of the Si substrate. Although BDD deposits on Nb support have been explored in several studies [46], their widespread use is likely impractical due to their prohibitively expensive cost. Since Ti has all the characteristics necessary to be a good and affordable substrate material, samples of tiny Ti/ BDD anodes have also been evaluated for the destruction of a number of dyes [47]. Since cracks develop and lead to the diamond film's separation during prolonged electrolysis, a synthetic method to deposit stable diamond films on Ti at an industrial scale is not yet available.

Dimensionally Stable Anode (DSA)-type Electrodes

DSAs made with blends of Ti, Ir, Ru, Sn, and/or Sb oxides exhibit remarkable mechanical and chemical resistance even at high current densities and in severely acidic conditions. They also have a high surface area. Due to their poor capacity to electrogenerate OH, these active anodes exhibit a constrained ability to oxidize dyestuffs. Under comparable conditions, the Ti-Ru-Sn ternary oxide electrode removes substantially less COD from a Methyl Red solution than non-active anodes like Si/BDD and Ti/PbO_2 and even another active anode like Pt [48]. A Ti/Sb_2O_5-SnO_2 anode was used by Chen *et al.* [49] to oxidize 25 to 30 cm^3 of solutions of various azo dyes with 2 g/dm_3 Na_2SO_4. With a stainless-steel cathode and electrodes measuring 25 mm, 24 mm, and 1.6 mm in size, they used the stirred undivided cell. 750 mg/dm^3 of Orange II underwent 98% colour removal but only with 27% COD removal and 16% current efficiency after the consumption of 6.25 Ah/dm^3 at 20 mA/cm^2, whereas 1500 mg/dm^3 of Reactive Red HE-3B underwent 95% colour removal with 13% COD decay and 11% current efficiency [50]. For the same oxidation of 1 g/dm^3 of 15 reactive dyes at pH 4.7-6.3 and 10 mA/cm^2 up to 2.4-4.0 Ah/dm^3, low mineralization with 26-47% COD abatement, current efficiency between 19 and 48%, and around 80% decolorization efficiency were also discovered [51]. If oxidation intermediates such aliphatic acids build and are slowly eliminated, as predicted by the same authors. These findings show that Methyl Red is mostly electrogenerated from water discharge on Ti/PbO_2 and Si/BDD surfaces.

The percentage of colour removal and COD decay for particular synthetic dye wastewaters while using monopolar and bipolar Fe or steel electrodes under ideal EC conditions. In addition to the electrolytic system, the pH of the solution, the retention time (tr), the rate of stirring or flow, and the applied current density (or cell voltage) all have a role in the removal of dyes. An unstirred Fe/steel tank reactor with electrodes of 50 mm, 50 mm, and 3 mm in dimension operating in batch mode was used to study the effects of initial pH, retention time, and current density as functions of decolorization efficiency for 250 cm^3 of 50 mg dm^3 solutions of the azo dye Basic Red 46 in NaCl (conductivity 8 mS cm1). At pH 5.5-8.5, the maximum amount of colour removal roughly 95% is accomplished. Most electrogenerated Fe^{3+} under these circumstances generates Fe(OH)$_3$ flocs, which can quickly remove the dye molecules through complexation or electrostatic attraction, followed by coagulation. At pH > 9.0, a portion of Fe(OH)$_3$ is solubilized as Fe(OH)$_4$, and less dye can be separated. In contrast, at pH 3.0, soluble Fe^{3+} is the major species and Fe(OH)$_3$ flocs are fairly weakly formed. The total electrolysis duration of the dye effluent in the reactor is another crucial quantity. The best decolorization efficiency of a 50 mg/dm^3 Basic Red 46 solution at pH 5.8 and 6 mA/cm^2 requires a minimum retention duration of 5 min. At this rate, enough Fe(OH)$_3$ flocs are formed to remove the dyestuff effectively [51].

Anode Materials for Specific Dye

It is preferred to employ stirred tank reactors with stirring rates between 150 and 250 rpm when operating in batch mode, even if the retention period is a function of operating factors including current density and dye concentration [52]. As floc aggregation is gradually favoured and coagulation becomes simpler, Daneshvar *et al*. [53] reported that colour removal was improved with increasing stirring rate up to 200 rpm during the EC treatment of 250 cm^3 of 50 mg dm^3 of the azo dye Acid Orange 7 in a Fe/Fe cell. However, increased stirring rates result in flocs degrading with time and a decline in the effectiveness of colour removal. The effluent flow rate controls the retention time in flow cells. Using a 450 cm^3 flow-through cell with two monopolar and three intermediate bipolar steel plate electrodes (11.0 cm, 11.4 cm in dimension), Mollah *et al*. [54] discovered that the colour removal of a 30 mg/dm^3 Acid Orange 7 solution containing 0.034 M NaCl of pH 8.4 decays from 99 to 93% at 16 mA/cm^2 and 25 °C when the flow rate varies from 21 to 36 dm^3/h. Then, in order to get acceptable decolorization efficiency, low flow rates with sufficiently extended retention durations are needed.

A recent publication of exciting research on the EC treatment of textile industry sulphur dye wastewater [55] was made. The original affluent contained 1487 mg/dm^3 of COD, 594 mg/dm^3 of total suspended solids (TSS), and 9458 units of colour, according to the American Dye Manufacturers Institute. The electrochemical cell was made up of a 4 dm^3 tank reactor with a total contact surface area of 0.37 m² and electrodes made up of two monopolar end steel plates and three intermediate bipolar steel plates. With the use of a 33-factorial design in batch operation mode without stirring, the initial pH, applied current, and retention time were all optimized. After consuming 900 C, the performance was at its best starting at pH 5.0, leading to decays for COD, TSS, and color of 76, 93, and 99%, respectively.

Electrochemical oxidation and EC have both been examined in comparison for the removal of various azo dyes. With batch mode Fe/Fe, polypyrrole/steel, and boron-doped diamond (BDD)/Cu cells, Lopes *et al.* [56] treated 100-300 cm³ of 350 mg/dm³. Direct Red 80 solutions in 5 g/dm³ Na_2SO_4 or NH_4NO_3. Despite the fact that 99–100% of the colour was removed in each case, the COD decay was only 46-49% with Fe and polypyrrole anodes but increased significantly to 87% with a BDD anode. Contrarily, the EC process had the lowest energy cost (1.34 kWh/m³ at 5 mA/cm²), whereas the electrochemical oxidation of polypyrrole and BDD produced substantially higher values (5.46 kWh/m³ at 2 mA/cm² and 6.65 kWh/m³ at 1.5 mA/cm², respectively). Using different monopolar Fe, graphite, and IrO_2-TaO_2-RuO_2-coated titanium rod electrodes in batch mode, Muthukumar *et al.* [57] observed a comparable response for the treatment of 2 dm³ of 100 mg/dm³ acid orange 10 solutions in 0.017 M NaCl at pH 6-9 in a stirred tank reactor. With the iron anode, the matching color and COD decays were 98 and 60%, respectively, whereas the graphite anode completely decontaminated the solution. Aniline and 1-amino-2-naphthol-6,8-disulphonic acid were produced as a result of the reductive breakage of the azo linkage, with the former staying in the aqueous phase and the latter co-precipitating with $Fe(OH)_3$. Hsing *et al.* [58] also compared the removal rates of acid orange 6 by O_3, AOPs such as O_3/UV, O_3/TiO_2/UV, and Fenton's reagent, as well as EC utilizing a flow-through cell with two monopolar and three bipolar iron plate electrodes in the continuous mode. They demonstrated that, although having a substantially lower ability to remove the acid orange 6 with TOC 35% than photocatalytic UV techniques, EC offers a very quick decolorization of the solution.

The use of EC in combination with other chemical or electrochemical methods to speed up dye removal has been covered by a number of authors. For acid orange 7, electrocoagulation was combined with electro-oxidation using a cell with three-phase, three-dimensional electrodes [59]. Other examples include electrocoagulation assisted by O_3 for Reactive Black 5 [60], electrocoagulation assisted by cobalt phosphomolybdate modified kaolin for Acid Red 18 [61], and electrocoagulation combined with AOPs like TiO_2/UV and O_3/UV for Reactive Red 2 [62].

In addition to polluting the environment, the buildup of dyestuffs and dye-containing wastewater also raises issues of science and aesthetics. Due to the refractory nature of the dye and the ongoing fluctuation of effluent characteristics, the use of a single treatment no longer satisfies regulatory standards. Numerous authors have also stated that because of today's stricter rules governing the disposal of dye wastewater, using a single treatment process is no longer practical. Furthermore, it is impossible to eliminate all the colors, heavy metals, and other chemicals found in industrial effluent with a single treatment [63 - 65]. Therefore, by combining the numerous treatments, the drawbacks of a single therapy strategy may be overcome. Recently, hybrid methods for dye removal have received more attention from researchers. The distinction between included and hybrid processes is essential because the two are commonly mislabelled. A procedure that enables the sequential performance of one or more treatments is included. On the other hand, a hybrid technique is a combination of several different remedy techniques. In essence, a hybrid approach can conserve resources and money by conducting multiple processes in a single vessel. There is currently a lot of interest in the use of hybrid materials or processes to improve outcomes, not only in the treatment of wastewater from the textile industry but also in the removal of other contaminants from municipal wastewater [66], wastewater treated with pig slurry [67], and wastewater from poultry slaughterhouses [68]. Both the conventional method and a hybrid process combination are used by researchers to apply the combinations more precisely for dye removal therapy. Additionally, a variety of physio-chemical techniques have been integrated with biological processes in several studies for the removal of dyes, as shown in Table **1**.

Table 1. Effluent treatment using electrochemical methods.

Type of Effluent	Composition	Process	Anode	Cathode	Reactor Type	Efficiency	Refs.
Textile	COD = 470, TSS = 68, pH = 8.8	AO	BDD	Zr	UR	85% TOC removal	[69]
Textile	COD = 5957, pH = 7.3	AO	DSA	SS	FTR	94%–99% of colour removal, COD no reported	[70]
Textile	COD = 1224, BOD_5 = 324, pH = 4.8, SO_4^{2-} = 38, Cl⁻ = 234	EF	Pt	Carbon Fibre	UR	75.2% COD removal	[71]
Industrial landfill leachate	COD = 17,100–18,400, pH = 9, Fe = 20, Cl⁻ = 52,300, NH_4^+ = 1200	AO	Carbon Plate	SS	UR	83% DOC removal	[72]

(Table 1) cont.....

Type of Effluent	Composition	Process	Anode	Cathode	Reactor Type	Efficiency	Refs.
Textile	COD = 2000 DOC = 485, pHv = v6.8, TSS = 230	EC	Fe or Al in monopolar connection	Fe or Al	UR	85% COD removal (Fe) and 77% (Al)	[73]
Tannery	COD = 9922–10,180, BOD_5 = 528, TSS = 445–530, pH = 3.7–4.3, Cl^- = 1239, Fe = 2–2.8	EC/AO/EF/PEF + PEF-UVA	BDD	Fe/BDD	UR	95% EC + PEF	[74]
Mixed wastewater from WWTP	COD = 1152, TOC = 274, BOD_5/COD = 0.43, TDS = 8370, pH = 7.2, Cl^- = 3691	EF with persulfate + BIO	Ti/Pt	Graphite felt	UR	60% of COD removal with EF persulfate Overall 94% COD removal	[75]
Pulp and paper	COD = 1450, BOD_5 = 350, TSS = 350, TDS = 1050, Cl^- = 325, pH = 6.72	Permanganate oxidation + ECP + Co_3/UV/peroxymonosulfate	Fe	Fe	UR	Overall 95% COD removal	[76]

CONCLUSION

Even if there are numerous cutting-edge approaches and concepts for environmental preservation, the problem of water pollution still exists today. This study looked at the benefits and restrictions of several physical, chemical, and biological methods for dye removal from wastewater. Due to the complexity of industrial wastewater, colour removal from a single treatment is challenging and time-consuming. Combinations of dye-containing effluents can be easily handled, and integrated/hybrid processes can reach extremely high colour removal efficiencies. However, putting this technology into practice at the industrial scale is a slow process that needs a large land area and is restricted by several factors such as the type of operation, concentration level, and chemical toxicity, among others. Electrochemical methods have been shown to be simple, dependable, and cost-effective at the lab and pilot scales. As a result, adjustments have been made to boost the effectiveness of present biological methods; the technique for successfully removing colours from water. The potentials, constraints, and future possibilities of these technologies are also explored in order to guarantee their long-term operation and stability. Reusing dyeing effluents that have been electrochemically cleaned is an exciting concept since it could result in significant water and salt savings. In Mediterranean countries, where river flow rates are low and salinity is an increasing environmental concern, this kind of research is very important. Although the foundations of electrochemistry are simple, as this overview has demonstrated, these methods can be used for a variety of textile processing.

REFERENCES

[1]   J. Wang, and L. Chu, "Irradiation treatment of pharmaceutical and personal care products (PPCPs) in water and wastewater: An overview", *Radiat. Phys. Chem.*, vol. 125, pp. 56-64, 2016.
[http://dx.doi.org/10.1016/j.radphyschem.2016.03.012]

[2]   M. Trojanowicz, "Removal of persistent organic pollutants (POPs) from waters and wastewaters by the use of ionizing radiation", *Sci. Total Environ.*, vol. 718, p. 134425, 2020.
[http://dx.doi.org/10.1016/j.scitotenv.2019.134425] [PMID: 31843309]

[3]   L. Nizzetto, M. Macleod, K. Borgå, A. Cabrerizo, J. Dachs, A.D. Guardo, D. Ghirardello, K.M. Hansen, A. Jarvis, A. Lindroth, B. Ludwig, D. Monteith, J.A. Perlinger, M. Scheringer, L. Schwendenmann, K.T. Semple, L.Y. Wick, G. Zhang, and K.C. Jones, "Past, present, and future controls on levels of persistent organic pollutants in the global environment", *Environ. Sci. Technol.*, vol. 44, no. 17, pp. 6526-6531, 2010.
[http://dx.doi.org/10.1021/es100178f] [PMID: 20604560]

[4]   S. Garcia-Segura, J.D. Ocon, and M.N. Chong, "Electrochemical oxidation remediation of real wastewater effluents: A review", *Process Saf. Environ. Prot.*, vol. 113, pp. 48-67, 2018.
[http://dx.doi.org/10.1016/j.psep.2017.09.014]

[5]   P.B. Tchounwou, C.G. Yedjou, A.K. Patlolla, and D.J. Sutton, "Heavy metal toxicity and the environment", In: *Molecular, Clinical and Environmental Toxicology.*, A. Luch, Ed., vol. Vol. 101. , 2012, pp. 133-164.
[http://dx.doi.org/10.1007/978-3-7643-8340-4_6]

[6]   M. Jaishankar, T. Tseten, N. Anbalagan, B.B. Mathew, and K.N. Beeregowda, "Toxicity, mechanism and health effects of some heavy metals", *Interdiscip. Toxicol.*, vol. 7, no. 2, pp. 60-72, 2014.
[http://dx.doi.org/10.2478/intox-2014-0009] [PMID: 26109881]

[7]   M.M. Bello, A.A. Abdul Raman, and M. Purushothaman, "Applications of fluidized bed reactors in wastewater treatment – A review of the major design and operational parameters", *J. Clean. Prod.*, vol. 141, pp. 1492-1514, 2017.
[http://dx.doi.org/10.1016/j.jclepro.2016.09.148]

[8]   P. Brian Chaplin, "Advantages, disadvantages, and future challenges of the use of electrochemical technologies for water and wastewater treatment", In: *Electrochemical Water and Wastewater Treatment.*, C.A. Martinez-Huitle, M.A. Rodrigo, O. Scialdone, Eds., Elsevier, 2018, pp. 451-494.
[http://dx.doi.org/10.1016/B978-0-12-813160-2.00017-1]

[9]   C.A. Martínez-Huitle, A. De Battisti, S. Ferro, S. Reyna, M. Cerro-López, and M.A. Quiro, "Removal of the pesticide methamidophos from aqueous solutions by electrooxidation using Pb/PbO$_2$, Ti/SnO$_2$, and Si/BDD electrodes", *Environ. Sci. Technol.*, vol. 42, no. 18, pp. 6929-6935, 2008.
[http://dx.doi.org/10.1021/es8008419] [PMID: 18853811]

[10]   M. Zhou, and J. He, "Degradation of azo dye by three clean advanced oxidation processes: Wet oxidation, electrochemical oxidation and wet electrochemical oxidation—A comparative study", *Electrochim. Acta,* vol. 53, no. 4, pp. 1902-1910, 2007.
[http://dx.doi.org/10.1016/j.electacta.2007.08.056]

[11]   A. Buthiyappan, A.R. Abdul Aziz, W.M.A. Wan Daud, and M. Wan, "Recent advances and prospects of catalytic advanced oxidation process in treating textile effluents", *Rev. Chem. Eng.*, vol. 32, no. 1, pp. 1-47, 2016.

[http://dx.doi.org/10.1515/revce-2015-0034]

[12] O.M.L. Alharbi, A.A. Basheer, R.A. Khattab, and I. Ali, "Health and environmental effects of persistent organic pollutants", *J. Mol. Liq.,* vol. 263, pp. 442-453, 2018.
[http://dx.doi.org/10.1016/j.molliq.2018.05.029]

[13] F.C. Moreira, R.A.R. Boaventura, E. Brillas, and V.J.P. Vilar, "Electrochemical advanced oxidation processes: A review on their application to synthetic and real wastewaters", *Appl. Catal. B,* vol. 202, pp. 217-261, 2017.
[http://dx.doi.org/10.1016/j.apcatb.2016.08.037]

[14] M.A. Oturan, "Electrochemical advanced oxidation technologies for removal of organic pollutants from water", *Environ. Sci. Pollut. Res. Int.,* vol. 21, no. 14, pp. 8333-8335, 2014.
[http://dx.doi.org/10.1007/s11356-014-2841-8] [PMID: 24723353]

[15] C.L.P.S. Zanta, P.A. Michaud, C. Comninellis, A.R. De Andrade, and J.F.C. Boodts, "Electrochemical oxidation of *p*-chlorophenol on SnO$_2$–Sb$_2$O$_5$ based anodes for wastewater treatment", *J. Appl. Electrochem.,* vol. 33, no. 12, pp. 1211-1215, 2003.
[http://dx.doi.org/10.1023/B:JACH.0000003863.13587.b7]

[16] M. Sun, X. Wang, L.R. Winter, Y. Zhao, W. Ma, T. Hedtke, J-H. Kim, and M. Elimelech, "Electrified membranes for water treatment applications", *ACS ES&T Engineering,* vol. 1, no. 4, pp. 725-752, 2021.
[http://dx.doi.org/10.1021/acsestengg.1c00015]

[17] K. Cho, and M.R. Hoffmann, "Urea degradation by electrochemically generated reactive chlorine species: Products and reaction pathways", *Environ. Sci. Technol.,* vol. 48, no. 19, pp. 11504-11511, 2014.
[http://dx.doi.org/10.1021/es5025405] [PMID: 25219459]

[18] M.A. Zazouli, and L.R. Kalankesh, "Removal of precursors and disinfection by-products (DBPs) by membrane filtration from water; a review", *J. Environ. Health Sci. Eng.,* vol. 15, no. 1, p. 25, 2017.
[http://dx.doi.org/10.1186/s40201-017-0285-z] [PMID: 29234499]

[19] K.S. Bharathi, and S.T. Ramesh, "Removal of dyes using agricultural waste as low-cost adsorbents: A review", *Appl. Water Sci.,* vol. 3, no. 4, pp. 773-790, 2013.
[http://dx.doi.org/10.1007/s13201-013-0117-y]

[20] S. Krishnan, H. Rawindran, C.M. Sinnathambi, and J.W. Lim, "Comparison of various advanced oxidation processes used in remediation of industrial wastewater laden with recalcitrant pollutants", *IOP Conf. Ser.: Mater. Sci. Eng.,* vol. vol.206, p. 012089, 2017.
[http://dx.doi.org/10.1088/1757-899X/206/1/012089]

[21] S. Feng, C. Xueming, G. Ping, and C. Guohua, "Electrochemical removal of fluoride ions from industrial wastewater", *Chem. Eng. Sci,* vol. 987, no. (3-6), pp. 987-993, 2003.

[22] E. Carlos, P. Barrera-Díaz, P. Balderas-Hernández, and B. Bryan, "Electrocoagulation: Fundamentals and prospectives", In: *Electrochemical Water and Wastewater Treatment.,* C.A. Martinez-Huitle, M.A. Rodrigo, O. Scialdone, Eds., , 2018, pp. 61-76.

[23] E. Butler, Y.T. Hung, R.Y.L. Yeh, and M. Suleiman Al Ahmad, "Electrocoagulation in wastewater treatment", *Water,* vol. 3, no. 2, pp. 495-525, 2011.
[http://dx.doi.org/10.3390/w3020495]

[24] P. Cañizares, F. Martínez, C. Jiménez, J. Lobato, and M.A. Rodrigo, "Coagulation and electrocoagulation of wastes polluted with dyes", *Environ. Sci. Technol.,* vol. 40, no. 20, pp. 6418-6424, 2006.
[http://dx.doi.org/10.1021/es0608390] [PMID: 17120574]

[25] D.T. Moussa, M.H. El-Naas, M. Nasser, and M.J. Al-Marri, "A comprehensive review of electrocoagulation for water treatment: Potentials and challenges", *J. Environ. Manage.,* vol. 186, no. Pt 1, pp. 24-41, 2017.

[http://dx.doi.org/10.1016/j.jenvman.2016.10.032] [PMID: 27836556]

[26]   C. Espinoza-Cisternas, and R. Salazar, "Application of electrochemical processes for treating effluents from landfill leachate as well as the agro and food industries", In: *Electrochemical Water and Wastewater Treatment.,* C.A. Martínez-Huitle, M.A. Rodrigo, O. Scialdone, Eds., Elsevier, 2018, pp. 393-419.
[http://dx.doi.org/10.1016/B978-0-12-813160-2.00015-8]

[27]   L. Szpyrkowicz, "Hydrodynamic effects on the performance of electro-coagulation/electro-flotation for the removal of dyes from textile wastewater", *Ind. Eng. Chem. Res.,* vol. 44, no. 20, pp. 7844-7853, 2005.
[http://dx.doi.org/10.1021/ie0503702]

[28]   W. Perren, A. Wojtasik, and Q. Cai, "Removal of microbeads from wastewater using electrocoagulation", *ACS Omega,* vol. 3, no. 3, pp. 3357-3364, 2018.
[http://dx.doi.org/10.1021/acsomega.7b02037] [PMID: 31458591]

[29]   J.B. Parsa, M. Rezaei, and A.R. Soleymani, "Electrochemical oxidation of an azo dye in aqueous media investigation of operational parameters and kinetics", *J. Hazard. Mater.,* vol. 168, no. 2-3, pp. 997-1003, 2009.
[http://dx.doi.org/10.1016/j.jhazmat.2009.02.134] [PMID: 19345003]

[30]   S.M.K.N. Islam, A.S.W. Kurny, and F. Gulshan, "Degradation of commercial dyes using mill scale by photo-fenton", *Environ. Process.,* vol. 2, no. 1, pp. 215-224, 2015.
[http://dx.doi.org/10.1007/s40710-014-0055-1]

[31]   D. Mohan, and C.U. Pittman Jr, "Activated carbons and low cost adsorbents for remediation of tri- and hexavalent chromium from water", *J. Hazard. Mater.,* vol. 137, no. 2, pp. 762-811, 2006.
[http://dx.doi.org/10.1016/j.jhazmat.2006.06.060] [PMID: 16904258]

[32]   L. Fan, Y. Zhou, W. Yang, G. Chen, and F. Yang, "Electrochemical degradation of aqueous solution of Amaranth azo dye on ACF under potentiostatic model", *Dyes Pigments,* vol. 76, no. 2, pp. 440-446, 2008.
[http://dx.doi.org/10.1016/j.dyepig.2006.09.013]

[33]   N.P. Shetti, S.J. Malode, R.S. Malladi, S.L. Nargund, S.S. Shukla, and T.M. Aminabhavi, "Electrochemical detection and degradation of textile dye Congo red at graphene oxide modified electrode", *Microchem. J.,* vol. 146, pp. 387-392, 2019.
[http://dx.doi.org/10.1016/j.microc.2019.01.033]

[34]   W. Zhou, Y. Ding, J. Gao, K. Kou, Y. Wang, X. Meng, S. Wu, and Y. Qin, "Green electrochemical modification of RVC foam electrode and improved H_2O_2 electrogeneration by applying pulsed current for pollutant removal", *Environ. Sci. Pollut. Res. Int.,* vol. 25, no. 6, pp. 6015-6025, 2018.
[http://dx.doi.org/10.1007/s11356-017-0810-8] [PMID: 29238928]

[35]   P.A. Carneiro, C.S. Fugivara, R F P. Nogueira, N. Boralle, M.V.B. Nivaldo, and B.Z. Maria, "A comparative study on chemical and electrochemical degradation of reactive blue 4 dye", *Port. Electrochem. Acta,* vol. 21, no. 1, pp. 49-67, 2003.
[http://dx.doi.org/10.4152/pea.200301049]

[36]   V. López-Grimau, and M.C. Gutiérrez, "Decolourisation of simulated reactive dyebath effluents by electrochemical oxidation assisted by UV light", *Chemosphere,* vol. 62, no. 1, pp. 106-112, 2006.
[http://dx.doi.org/10.1016/j.chemosphere.2005.03.076] [PMID: 15893798]

[37]   A.G. Vlyssides, M. Loizidou, P.K. Karlis, A.A. Zorpas, and D. Papaioannou, "Electrochemical oxidation of a textile dye wastewater using a Pt/Ti electrode", *J. Hazard. Mater.,* vol. 70, no. 1-2, pp. 41-52, 1999.
[http://dx.doi.org/10.1016/S0304-3894(99)00130-2] [PMID: 10611427]

[38]   L. Szpyrkowicz, R. Cherbanski, and G.H. Kelsall, "Hydrodynamic effects on the performance of an electrochemical reactor for destruction of disperse dyes", *Ind. Eng. Chem. Res.,* vol. 44, no. 7, pp. 2058-2068, 2005.

[http://dx.doi.org/10.1021/ie049444k]

[39]   L. Szpyrkowicz, M. Radaelli, S. Daniele, A. Baldacci, and S. Kaul, "Application of electro-catalytic mediated oxidation for the treatment of a spent textile bath in a membrane reactor", *Ind. Eng. Chem. Res.,* vol. 46, no. 21, pp. 6732-6736, 2007.
[http://dx.doi.org/10.1021/ie0616388]

[40]   C. Cameselle, M. Pazos, and M.A. Sanromán, "Selection of an electrolyte to enhance the electrochemical decolourisation of indigo. Optimisation and scale-up", *Chemosphere,* vol. 60, no. 8, pp. 1080-1086, 2005.
[http://dx.doi.org/10.1016/j.chemosphere.2005.01.018] [PMID: 15993155]

[41]   C.T. Wang, "Decolorization of Congo Red with three-dimensional flow-by packed-bed electrodes", *J. Environ. Sci. Health Part A Tox. Hazard. Subst. Environ. Eng.,* vol. 38, no. 2, pp. 399-413, 2003.
[http://dx.doi.org/10.1081/ESE-120016903] [PMID: 12638704]

[42]   L. Szpyrkowicz, C. Juzzolino, S.N. Kaul, S. Daniele, and M.D. De Faveri, "Electrochemical oxidation of dyeing baths bearing disperse dyes", *Ind. Eng. Chem. Res.,* vol. 39, no. 9, pp. 3241-3248, 2000.
[http://dx.doi.org/10.1021/ie9908480]

[43]   C.A. Martínez-Huitle, E.V. dos Santos, D.M. de Araújo, and M. Panizza, "Applicability of diamond electrode/anode to the electrochemical treatment of a real textile effluent", *J. Electroanal. Chem.,* vol. 674, pp. 103-107, 2012.
[http://dx.doi.org/10.1016/j.jelechem.2012.02.005]

[44]   D. Clematis, and M. Panizza, "Application of boron-doped diamond electrodes for electrochemical oxidation of real wastewaters", *Curr. Opin. Electrochem.,* vol. 30, p. 100844, 2021.
[http://dx.doi.org/10.1016/j.coelec.2021.100844]

[45]   Y. Kong, Z. Wang, Y. Wang, J. Yuan, and Z. Chen, "Degradation of methyl orange in artificial wastewater through electrochemical oxidation using exfoliated graphite electrode", *N. Carbon Mater.,* vol. 26, no. 6, pp. 459-464, 2011.
[http://dx.doi.org/10.1016/S1872-5805(11)60092-9]

[46]   A.S. Koparal, Y. Yavuz, C. Gürel, and Ü.B. Öğütveren, "Electrochemical degradation and toxicity reduction of C.I. Basic Red 29 solution and textile wastewater by using diamond anode", *J. Hazard. Mater.,* vol. 145, no. 1-2, pp. 100-108, 2007.
[http://dx.doi.org/10.1016/j.jhazmat.2006.10.090] [PMID: 17140728]

[47]   M. Cerón-Rivera, M.M. Dávila-Jiménez, and M.P. Elizalde-González, "Degradation of the textile dyes Basic yellow 28 and Reactive black 5 using diamond and metal alloys electrodes", *Chemosphere,* vol. 55, no. 1, pp. 1-10, 2004.
[http://dx.doi.org/10.1016/j.chemosphere.2003.10.060] [PMID: 14720540]

[48]   Y. Jiang, H. Zhao, J. Liang, L. Yue, T. Li, Y. Luo, Q. Liu, S. Lu, A.M. Asiri, Z. Gong, and X. Sun, "Anodic oxidation for the degradation of organic pollutants: Anode materials, operating conditions and mechanisms. A mini review", *Electrochem. Commun.,* vol. 123, p. 106912, 2021.
[http://dx.doi.org/10.1016/j.elecom.2020.106912]

[49]   X. Chen, G. Chen, and P.L. Yue, "Anodic oxidation of dyes at novel Ti/B-diamond electrodes", *Chem. Eng. Sci.,* vol. 58, no. 3-6, pp. 995-1001, 2003.
[http://dx.doi.org/10.1016/S0009-2509(02)00640-1]

[50]   X. Chen, and G. Chen, "Anodic oxidation of Orange II on Ti/BDD electrode: Variable effects", *Separ. Purif. Tech.,* vol. 48, no. 1, pp. 45-49, 2006.
[http://dx.doi.org/10.1016/j.seppur.2005.07.024]

[51]   X. Chen, F. Gao, and G. Chen, "Comparison of Ti/BDD and Ti/SnO$_2$?Sb$_2$O$_5$ electrodes for pollutant oxidation", *J. Appl. Electrochem.,* vol. 35, no. 2, pp. 185-191, 2005.
[http://dx.doi.org/10.1007/s10800-004-6068-0]

[52]   N. Daneshvar, A.R. Khataee, A.R. Amani Ghadim, and M.H. Rasoulifard, "Decolorization of C.I.

Acid Yellow 23 solution by electrocoagulation process: Investigation of operational parameters and evaluation of specific electrical energy consumption (SEEC)", *J. Hazard. Mater.,* vol. 148, no. 3, pp. 566-572, 2007.
[http://dx.doi.org/10.1016/j.jhazmat.2007.03.028] [PMID: 17428605]

[53]   N. Daneshvar, H. Ashassi-Sorkhabi, and A. Tizpar, "Decolorization of orange II by electrocoagulation method", *Separ. Purif. Tech.,* vol. 31, no. 2, pp. 153-162, 2003.
[http://dx.doi.org/10.1016/S1383-5866(02)00178-8]

[54]   M. Mollah, S.R. Pathak, P.K. Patil, M. Vayuvegula, T.S. Agrawal, J.A. Gomes, M. Kesmez, and D.L. Cocke, "Treatment of orange II azo-dye by electrocoagulation (EC) technique in a continuous flow cell using sacrificial iron electrodes", *J. Hazard. Mater.,* vol. 109, no. 1-3, pp. 165-171, 2004.
[http://dx.doi.org/10.1016/j.jhazmat.2004.03.011] [PMID: 15177756]

[55]   P. Fongsatitkul, P. Elefsiniotis, and B. Boonyanitchakul, "Treatment of a textile dye wastewater by an electrochemical process", *J. Environ. Sci. Health Part A Tox. Hazard. Subst. Environ. Eng.,* vol. 41, no. 7, pp. 1183-1195, 2006.
[http://dx.doi.org/10.1080/10934520600656372] [PMID: 16854794]

[56]   A. Lopes, S. Martins, A. Morão, M. Magrinho, and I. Gonçalves, "Degradation of a Textile Dye C. I. Direct red 80 by electrochemical processes", *Port. Electrochem. Acta,* vol. 22, no. 3, pp. 279-294, 2004.
[http://dx.doi.org/10.4152/pea.200403279]

[57]   M. Muthukumar, M.T. Karuppiah, and G.B. Raju, "Electrochemical removal of CI Acid orange 10 from aqueous solutions", *Separ. Purif. Tech.,* vol. 55, no. 2, pp. 198-205, 2007.
[http://dx.doi.org/10.1016/j.seppur.2006.11.014]

[58]   H.J. Hsing, P.C. Chiang, E.E. Chang, and M.Y. Chen, "The decolorization and mineralization of Acid Orange 6 azo dye in aqueous solution by advanced oxidation processes: A comparative study", *J. Hazard. Mater.,* vol. 141, no. 1, pp. 8-16, 2007.
[http://dx.doi.org/10.1016/j.jhazmat.2006.05.122] [PMID: 17222965]

[59]   Y. Xiong, P.J. Strunk, H. Xia, X. Zhu, and H.T. Karlsson, "Treatment of dye wastewater containing acid orange II using a cell with three-phase three-dimensional electrode", *Water Res.,* vol. 35, no. 17, pp. 4226-4230, 2001.
[http://dx.doi.org/10.1016/S0043-1354(01)00147-6] [PMID: 11791853]

[60]   S. Song, Z. He, J. Qiu, L. Xu, and J. Chen, "Ozone assisted electrocoagulation for decolorization of C.I. Reactive Black 5 in aqueous solution: An investigation of the effect of operational parameters", *Separ. Purif. Tech.,* vol. 55, no. 2, pp. 238-245, 2007.
[http://dx.doi.org/10.1016/j.seppur.2006.12.013]

[61]   Q. Zhuo, H. Ma, B. Wang, and L. Gu, "Catalytic decolorization of azo-stuff with electro-coagulation method assisted by cobalt phosphomolybdate modified kaolin", *J. Hazard. Mater.,* vol. 142, no. 1-2, pp. 81-87, 2007.
[http://dx.doi.org/10.1016/j.jhazmat.2006.07.063] [PMID: 17005320]

[62]   C-H. Wu, C-L. Chang, and C-Y. Kuo, "Decolorization of Procion Red MX-5B in electrocoagulation (EC), UV/TiO$_2$ and ozone-related systems", *Dyes Pigments,* vol. 76, no. 1, pp. 187-194, 2008.
[http://dx.doi.org/10.1016/j.dyepig.2006.08.017]

[63]   F.I. Hai, K. Yamamoto, and K. Fukushi, "Hybrid treatment systems for dye wastewater", *Crit. Rev. Environ. Sci. Technol.,* vol. 37, no. 4, pp. 315-377, 2007.
[http://dx.doi.org/10.1080/10643380601174723]

[64]   S.R. Geed, K. Samal, and A. Tagade, "Development of adsorption-biodegradation hybrid process for removal of methylene blue from wastewater", *J. Environ. Chem. Eng.,* vol. 7, no. 6, p. 103439, 2019.
[http://dx.doi.org/10.1016/j.jece.2019.103439]

[65]   W. Li, B. Mu, and Y. Yang, "Feasibility of industrial-scale treatment of dye wastewater *via* bio-adsorption technology", *Bioresour. Technol.,* vol. 277, pp. 157-170, 2019.

[http://dx.doi.org/10.1016/j.biortech.2019.01.002] [PMID: 30638884]

[66]   T. Agarwal, R. Narayan, S. Maji, S. Behera, S. Kulanthaivel, T.K. Maiti, I. Banerjee, K. Pal, and S. Giri, "Gelatin/Carboxymethyl chitosan based scaffolds for dermal tissue engineering applications", *Int. J. Biol. Macromol.*, vol. 93, no. Pt B, pp. 1499-1506, 2016.
[http://dx.doi.org/10.1016/j.ijbiomac.2016.04.028] [PMID: 27086289]

[67]   F.F. Vinicius, "Turning water abundance into sustainability in brazil", *Front. Environ. Sci.*, no. Dec, p. 727051, 2021.

[68]   C. Balcik-Canbolat, T. Olmez-Hanci, C. Sengezer, H. Sakar, A. Karagunduz, and B. Keskinler, "A combined treatment approach for dye and sulfate rich textile nanofiltration membrane concentrate", *J. Water Process Eng.*, vol. 32, p. 100919, 2019.
[http://dx.doi.org/10.1016/j.jwpe.2019.100919]

[69]   E. Tsantaki, T. Velegraki, A. Katsaounis, and D. Mantzavinos, "Anodic oxidation of textile dyehouse effluents on boron-doped diamond electrode", *J. Hazard. Mater.*, vol. 207-208, pp. 91-96, 2012.
[http://dx.doi.org/10.1016/j.jhazmat.2011.03.107] [PMID: 21530081]

[70]   S.S. Vaghela, A.D. Jethva, B.B. Mehta, S.P. Dave, S. Adimurthy, and G. Ramachandraiah, "Laboratory studies of electrochemical treatment of industrial azo dye effluent", *Environ. Sci. Technol.*, vol. 39, no. 8, pp. 2848-2855, 2005.
[http://dx.doi.org/10.1021/es035370c] [PMID: 15884385]

[71]   C-T. Wang, W-L. Chou, M-H. Chung, and Y-M. Kuo, "COD removal from real dyeing wastewater by electro-Fenton technology using an activated carbon fiber cathode", *Desalination*, vol. 253, no. 1-3, pp. 129-134, 2010.
[http://dx.doi.org/10.1016/j.desal.2009.11.020]

[72]   N. Nageswara Rao, M. Rohit, G. Nitin, P.N. Parameswaran, and J.K. Astik, "Kinetics of electrooxidation of landfill leachate in a three-dimensional carbon bed electrochemical reactor", *Chemosphere*, vol. 76, no. 9, pp. 1206-1212, 2009.
[http://dx.doi.org/10.1016/j.chemosphere.2009.06.009] [PMID: 19564036]

[73]   M. Kobya, E. Gengec, and E. Demirbas, "Operating parameters and costs assessments of a real dyehouse wastewater effluent treated by a continuous electrocoagulation process", *Chem. Eng. Process.*, vol. 101, pp. 87-100, 2016.
[http://dx.doi.org/10.1016/j.cep.2015.11.012]

[74]   E. Isarain-Chávez, C. de la Rosa, L.A. Godínez, E. Brillas, and J.M. Peralta-Hernández, "Comparative study of electrochemical water treatment processes for a tannery wastewater effluent", *J. Electroanal. Chem.*, vol. 713, pp. 62-69, 2014.
[http://dx.doi.org/10.1016/j.jelechem.2013.11.016]

[75]   A. Popat, P.V. Nidheesh, T.S. Anantha Singh, and M. Suresh Kumar, "Mixed industrial wastewater treatment by combined electrochemical advanced oxidation and biological processes", *Chemosphere*, vol. 237, p. 124419, 2019.
[http://dx.doi.org/10.1016/j.chemosphere.2019.124419] [PMID: 31356998]

[76]   N. Jaafarzadeh, F. Ghanbari, M. Ahmadi, and M. Omidinasab, "Efficient integrated processes for pulp and paper wastewater treatment and phytotoxicity reduction: Permanganate, electro-Fenton and Co_3O_4/UV/peroxymonosulfate", *Chem. Eng. J.*, vol. 308, pp. 142-150, 2017.
[http://dx.doi.org/10.1016/j.cej.2016.09.015]

CHAPTER 5

Effect of Electrode Materials in Decolorization of Dyestuffs from Wastewater

R. Jagatheesan[1,*], C. Christopher[2] and **K. Govindan[3]**

[1] *Department of Chemistry, Vivekanandha College of Arts and Sciences for Women (Autonomous), Elayampalayam, Tiruchengode, Tamil Nadu-637 205, India*

[2] *Department of Chemistry, St. Xavier's College, Palayamkottai, Tirunelveli-627 002, India*

[3] *Environmental System Laboratory, Department of Civil Engineering, Kyung Hee University (Global Campus), Giheung-Gu, Yongin-Si, Gyeonggi-Do-16705, Republic of Korea*

Abstract: The wastewater produced by the textile industry is replete with numerous contaminants that are known to be hazardous to aquatic and terrestrial living systems. Particularly dangerous contaminants in the textile sector that defy traditional degrading techniques include synthetic dyestuffs. In order to protect the environment, this chapter reviews current advancements in the electrochemical treatment of wastewater containing synthetic organic dyes by anodic oxidation. The mechanisms of electrochemical oxidation in anodic oxidation processes are thoroughly described. The electrochemical degradation of wastewater has been studied using a wide variety of electrodes. As a result, this paper attempts to summarize and discuss the most significant and recent studies on the use of anodes for the removal of organic synthetic dyestuffs that are currently available in the literature.

Keywords: Anodic oxidation, Decolorization, Dyestuffs, Organic dyes.

INTRODUCTION

It is inevitable that dyes will play a role, ideally in sectors like the textile, food, paper, and plastic industries. The toxins produced by these sectors provide a variety of life-threatening problems for the environment, which in turn severely disrupts aquatic life [1 - 4]. The water clarity, aesthetic appeal, and gas solubility are all significantly impacted by dye concentrations, even at very low levels [5]. Because of this, dyes are carcinogenic, poisonous, and teratogenic, and the process of photochemical and biochemical degradation is highly difficult for complete mineralization as secondary contaminants are produced. Additionally, this contaminated wastewater is not recommended for human consumption or

---

* **Corresponding author R. Jagatheesan:** Department of Chemistry, Vivekanandha College of Arts and Sciences for Women (Autonomous), Elayampalayam, Tiruchengode, Tamil Nadu-637 205, India; E-mail: jagan3311d@gmail.com

Paulpandian Muthu Mareeswaran & Jegathalaprathaban Rajesh (Eds.)

even for domestic use [6]. Critical studies have so far reported on a variety of techniques, such as physio-chemical, chemical, and biological oxidation, for removing colours from dye effluents [7 - 10]. The method is costly, less successful in removing colour, and less favourable to a variety of dye-polluted waterways. Anodic electrochemical oxidation has received a lot of attention in recent years due to its low cost, lack of sludge creation, simple operating procedure, and ecologically friendly process. Additionally, the organic matter is intended to be mineralized into CO_2 or changed into biocompatible chemicals by this process [11]. Thus, because the electrode is so important to the electrochemical oxidation process, novel anode materials are increasingly being preferred to boost current efficiency [12]. To lay the groundwork for the electrochemical treatment of dyestuffs, we outlined in this chapter potential dye degradation procedures employing different anodic electrode materials.

Oxidation Mechanisms

Electrochemical cell types (divided/undivided), electrolyte medium, electrode materials, and power supply are the fundamental requirements for bulk electrolysis. Typically, no separator is used in the undivided batch electrochemical redox tests (Scheme **1**). However, divided cells have a membrane-based divider that separates the anodic and cathodic sections, and reactions happen in several compartments (Scheme **2**). A semi-porous membrane divides the cathode and anode chambers into divided cells. Sintered glass, porous porcelain, polytetrafluoroethylene, and polypropylene are examples of typical membrane materials. The function of divided cell is to allow ion diffusion while limiting the flow of products and reactants. Galvanostatic conditions in an undivided cell setup using direct or indirect electrochemical electron transfer reactions are taken into consideration for the large-scale setup in the dye degradation process (Table **1**) [13].

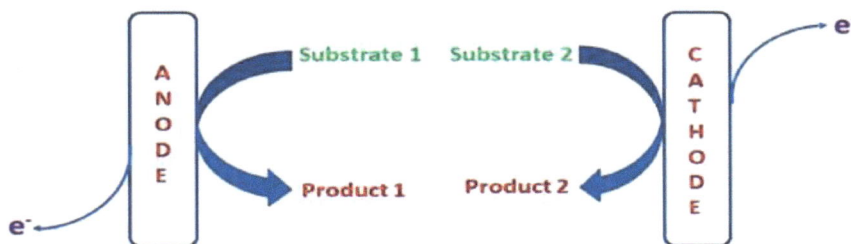

Scheme (1). Schematic electrolysis in an undivided cell.

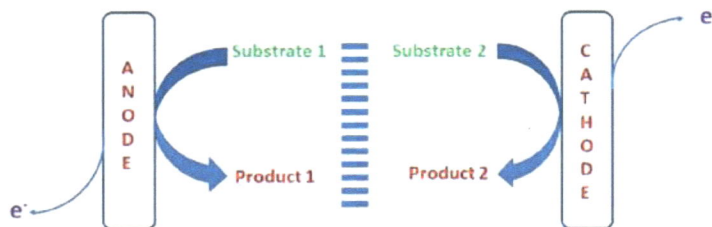

Scheme (2). Schematic electrolysis in a divided cell.

Table 1. Direct or indirect electrochemical oxidation reaction-Summary [13, 14].

S. No.	Direct Oxidation	Indirect Oxidation
1		
2	$R \xrightarrow[\text{Anode}]{-e} R^{\bullet+}$	$M \xrightarrow[\text{Anode}]{-e} M^{+} \xrightarrow{R} R^{\bullet-} + M$
3	The electron exchange takes place directly between the electrode and the pollutants.	There is no direct electron exchange between the organic contaminants and the anode surface, but the primary electron exchange *via* an oxidant*
4	Heterogeneous reaction	Homogeneous/heterogeneous
5	Adsorption of the pollutant on the surface leads to the formation of a polymer layer on the anode surface (passivation).	The pollutant's adsorption must be minimized as much as possible.
6	The degradation of organic substances is not very effective.	Efficient process for the degradation of organic pollutants.

* Strong oxidizing agents or metallic redox couples can act as oxidation mediators.

Metallic Redox Couple- Ag (II), Co(III), Ce (IV), and Fe (III)

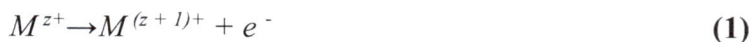

$$M^{z+} \rightarrow M^{(z+1)+} + e^{-} \tag{1}$$

$$M^{(z+1)+} + organics \rightarrow M^{z+} + CO_2 + H_2O \tag{2}$$

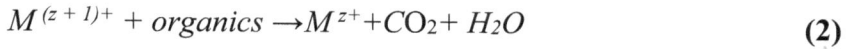

Mediated Electrochemical Oxidation (MEO) is the name given to the indirect electrolysis caused by metallic couples acting as redox reagents [14]. The metal ions undergo anodical oxidation in acidic conditions, transitioning from their stable oxidation state (M^{z+}) to the more reactive oxidation state ($M^{(z+1)+}$), where they attack and decompose other organic molecules into carbon dioxide, insoluble inorganic salts, and water [15].

Oxidizing chemicals - Active chlorine

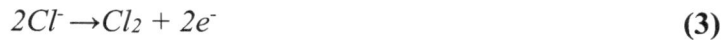

$$2Cl^- \rightarrow Cl_2 + 2e^- \tag{3}$$

By generating *in-situ* strongly oxidizing agents, organic pollutants can be degraded by indirect electrolysis, -persulfate, perphosphate or percarbonate.

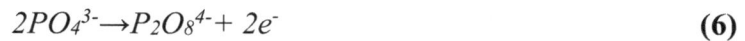

$$2SO_4^{2-} \rightarrow S_2O_8^{2-} + 2e^- \tag{4}$$

$$2CO_3^{2-} \rightarrow C_2O_6^{2-} + 2e^- \tag{5}$$

$$2PO_4^{3-} \rightarrow P_2O_8^{4-} + 2e^- \tag{6}$$

These compounds are acquired by anodic oxidation of SO_4^{2-}, CO_3^{2-} and $2PO_4^{3-}$ present in the solution as above [15].

Oxidation Involving Intermediates of Oxygen Evolution

In general, organic pollutants undergo degradation at high anodic potentials in the water discharge zone due to the association of oxygen evolution intermediates [16].

$$M + H_2O \rightarrow M(^{\bullet}OH) + H^+ + e^- \tag{7}$$

$$R + M(^{\bullet}OH) \rightarrow M + CO_2 + H_2O \tag{8}$$

The anodic oxidation of aqueous effluent usually follows two pathways:

Electrochemical conversion: The contaminants need additional biological conversion because they have already been partially transformed into more biodegradable reaction by-products.

Electrochemical combustion: This technique is exceptional in that it treats organic contaminants without the need for additional purification. In particular, it transforms the original organic contaminants into less dangerous tiny molecules like water, carbon dioxide, and other inorganic species [17 - 19].

Electrochemical degradation / decolorization of dyestuffs from wastewater: The wastewater from the textile industry is released, and it contains a variety of organic contaminants that are recognized to be hazardous for both aquatic and terrestrial life. Many researchers have experimented with various strategies to degrade the dyestuffs from wastewater to solve these issues. We have succinctly outlined some recent efforts to electrochemically degrade dyestuffs from industrial effluent in this chapter.

Reactive - Azo Dyes

Reactive Orange - RO16

BDD/Ti electrodes were used to treat the water that had been tainted by Reactive Orange - RO16 (Fig. **1**) [20]. Using the hot filament chemical vapor deposition (HFCVD) method, two different doping concentration levels were added to the electrodes. The E1 and E2 electrodes, with concentrations of 4.0 and 8.0 1021 atoms/cm^3, respectively, were used as growth parameters to produce heavily doped diamond films. The film quality and its physical characteristics were assessed using Mott-Schottky plots and Raman spectra. Studies using scanning electron microscopy showed that the crystal orientation combinations of (1 1 1) and (1 0 0) were present in well-defined microcrystalline grain morphologies. The most doped electrode (E2) demonstrated outstanding efficiency in case of the aromatic moiety reduction and the azo group degradation from the UV-visible spectra. The total organic carbon (TOC) and HPLC chromatographic results also supported these patterns. Finally, the findings showed that there is a direct correlation between the BDD morphology, physical characteristics, and performance in the Reactive Orange - RO16 degrading process.

Reactive Orange 16, Reactive Violet 4, Reactive Red 228, and Reactive Black 5

Reactive Orange 16, Reactive Violet 4, Reactive Red 228, and Reactive Black 5 (Figs. **2**, **3** and **4**) are some of the unique instances of numerous dyes that have particular functional groups and the same molecular basic structure and are used for an electrochemical decolorization investigation. The functional group of the selected dye largely lowers the nucleophilicity of the contaminant and prevents the electrophilic assault of the electrogenerated hydroxyl radical (•OH). Thus, both the pollutant's rate of decolorization and the total decolorization efficiency are significantly decreased. The double bond conjugation of the pollutant selectively increases the activation energy of the system needed for the electrophilic attack of •OH, which primarily affects the efficacy of the electrochemical technology in treating wastewater [21]. In addition, the N=N bond chromophore present in the dye increases the recalcitrant phenomena.

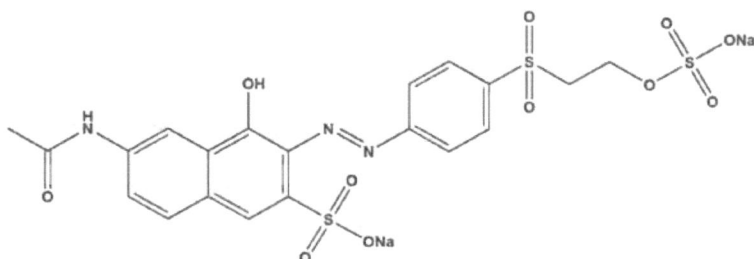

Fig. (1). Structure of reactive orange 16.

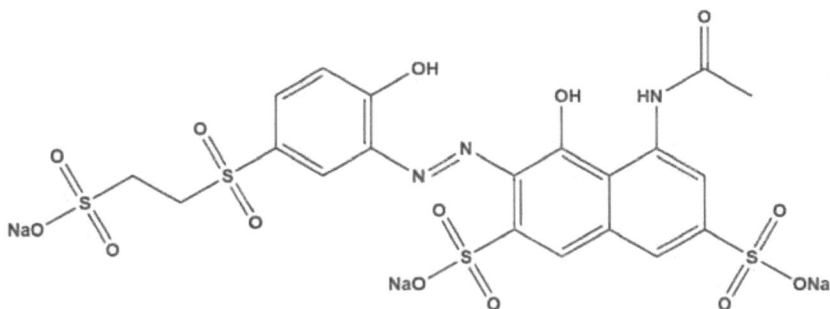

Fig. (2). Structure of reactive violet 4.

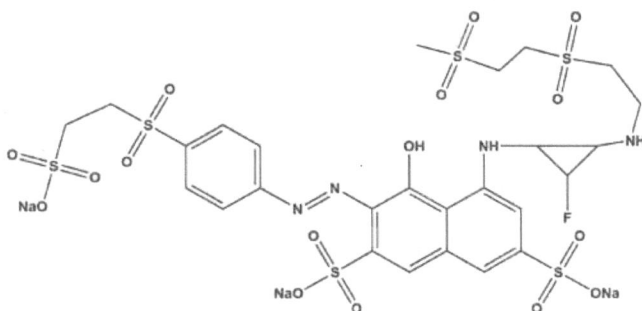

Fig. (3). Structure of Reactive Red 228.

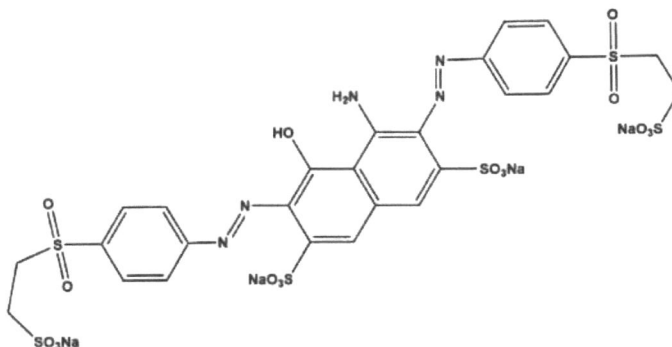

Fig. (4). Structure of Reactive Black 5.

The present study focuses on the deduction of a recalcitrant and toxic dye, Reactive Black 5 (RB-5), by three batch modes of electrochemical operation. They are:

i) Anodic oxidation (AO) of dye on the BDD electrode.

ii) Reduction of oxygen using reticulated vitreous carbon (RVC) electrodes in the Electro- Fenton (EF) process involving oxidant, and hydrogen peroxide.

Hydrogen peroxide is used as an oxidant in the Electro-Fenton (EF) process.

A Nafion membrane-separated division cell has been used to do the procedures. The contaminants were placed in both the anolyte and the catholyte compartments in this experimental setup. Additionally, this setup proved more effective than the combined approach at removing colour and significant amounts of total organic carbon (TOC), ranging from 74 to 82 percent. We assessed the effects of initial Fe^{2+} ion concentration and current density. When 0.4 V vs Ag/AgCl was given to an RVC electrode and the concentration of Fe^{2+} was 1.0 104 mol/dm³, and total colour and 74% of TOC reductions were achieved in less than 90 minutes of electrolysis, yielding the lowest energy effectiveness (208 kWh/kg) with the EF process. The BDD electrode had the highest apparent rate constant (0.835 min^{-1}) and was the best material for removing the RB-5 dye in 7.5 minutes. It also removed 82% of the TOC in 30 minutes while using 291 kWh/kg^{-1} of energy and 41.1 mA/cm^{-2} of current density. Hence, it was more efficient to apply the oxidation process separately.

iii) The combination of AO and EF methods – This method makes use of beaker-style cells. Here, the electrodegradation rate of RB-5 was at least three times lower, with a notable rate constant (0.269 min^{-1}) and a high EC (682 kWh/kg^{-1}) that removed 32% of the TOC [22].

Reactive Yellow 135

Both decolorization and electrocoagulation (EC) were individually shown in this wastewater treatment procedure. The beginning pH, current density, electrolysis time, and initial dye concentrations of the decolorization procedure were all tuned using Reactive Yellow 135 (Fig. **5**). Second, in the EC process, Sacrificial Anodes (Al) were used, and the amount of sludge created in the same was also reported per m³ of pollutant. Separate chemical coagulation studies utilizing aluminium salts were also carried out to determine the efficacy of the EC procedure.

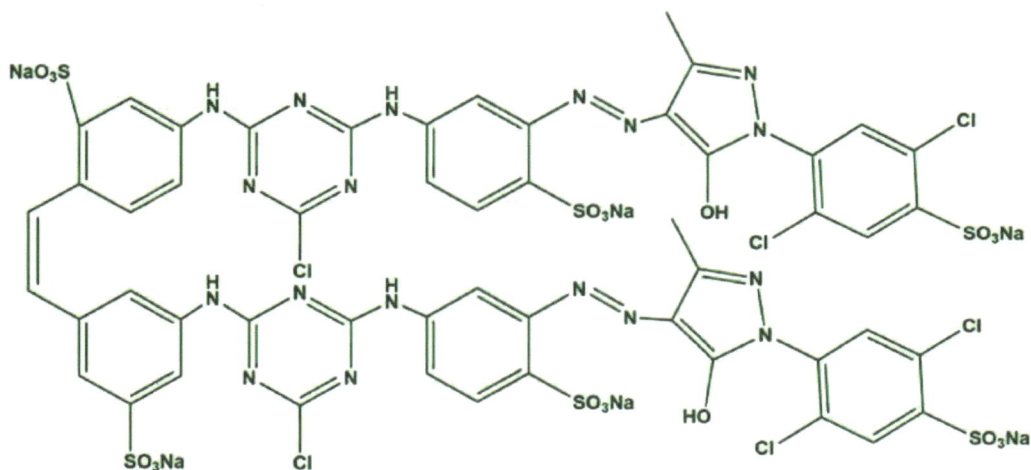

Fig. (5). Chemical structures of Reactive Yellow 135.

The treatment performance was assessed using dye removal effectiveness and total organic carbon (TOC). This study demonstrated that treating textile wastewater with EC successfully reduced COD, TOC, turbidity, and TSS by 81%, 85%, 93.7%, and 97.1% while maintaining a low operating cost. The method required 0.11 kWh/m³ of energy, and 0.03 kg/m³ of aluminum, and produced 0.15 kg/m³ of sludge [23].

Sunset Yellow-SY

Sunset yellow-SY (Fig. **6**) azo dye (100 mL of 290 mg L1) decolorization and mineralization were demonstrated at the ideal pH 3 using a BDD anode and a carbon-PTFE air-diffusion cathode in an undivided cell. Anodic oxidation, electro-Fenton (EF), UV-assisted photoelectro-Fenton (PEF), and solar photoelectro-Fenton (SPEF) were used to carry out this procedure. Here, the hydroxyl radical (•OH) produced by the processes was used to eliminate organic materials. Due to the increased UV intensity of sunshine, which quickly photolyzes Fe^{3+}carboxylate complexes that cannot be eliminated by •OH in EF, SPEF was more effective in attaining an almost entire mineralization more quickly than PEF. Nevertheless, SY was totally decoloured by EF, PEF, and SPEF at comparable rates. The mild oxidation activity of •OH at the BDD anode led to a gradual decolorization and mineralization in AO-H_2O_2. In general, the pseudo-first-order reaction was chosen by the azo dye degradation. It shows that decolorization took place more slowly than the elimination of contaminants. However, the decolorization process includes colourful aromatic compounds. Around 14 aromatic products and 34 hydroxylated derivatives (including phthalic acid compounds, naphthalenic compounds, and benzenic compounds) were found using the LC-MS approach. Using the ion-exclusion HPLC approach, generated carboxylic acids (tartronic, oxalic, formic, and oxamic) were also detected. To demonstrate the practicality of SPEF on an industrial scale, a solar pre-pilot plant equipped with a Pt/carbon-PTFE air-diffusion cell and a compound parabolic collectors (CPCs) photoreactor was deployed. In this facility, 100% SY decolorization was achieved between 33.2 and 77.6 mA/cm², and mineralization was achieved between 91 and 94% in a very short period [24].

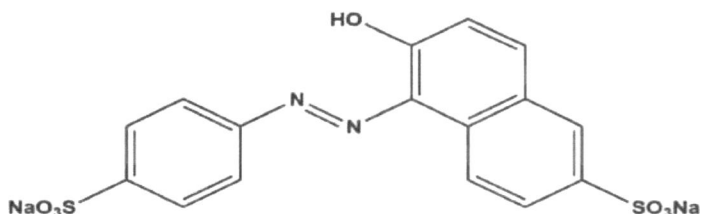

Fig. (6). Structural formula of Sunset Yellow FCF (SY) azo dye.

Utilizing Fe-C micro-electrolysis under the ideal conditions of starting pH 6, Fe/C ratio 1:1, Fe/C total mass 500 gL^{-1}, and airflow rate 45 Lh^{-1}, another technique of SY wastewater degradation was developed. The situation examines how the degradation and COD removal efficiency of SY were roughly 99.0% and 77.5% after 90 min of treatment. 36 mm spherical coal-based activated carbon was chosen as the ideal electrode material. Additionally, the efficacy of the Fe-C micro-electrolysis technique for degrading additional refractory organic wastewaters was examined using the antibiotic and printing wastewaters. SEM, XPS, and XRD were used to characterize the AC surfaces before and after 20 times of experimentation. Additionally, UV-Vis analysis was used to determine the disintegration and degradation process of the treated SY wastewaters [25].

Alizarin Red S

Using galvanostatic and potentiostatic techniques, the electrooxidation of the anthraquinone dye Alizarin Red S (Fig. **7**) was studied on BDD electrodes in an acidic solution. The bulk electrolysis solution's measured COD and TOC provide evidence that organic materials were completely converted into carbon dioxide through a series of intricate processes. The organic chemicals in wastewater have now been eliminated at both low and high current densities. When mass-transfer occurs at the anode surface as opposed to the mediated process by •OH and peroxodisulfate, this approach has limitations. As a result, this electrochemical method is ideal for treating industrial wastewaters containing anthraquinone dyes since it is highly clean [26].

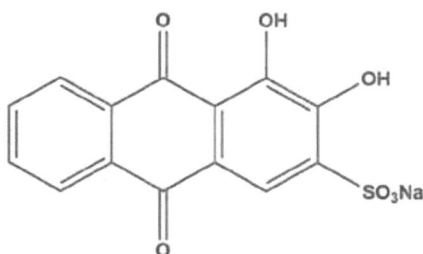

Fig. (7). Chemical structure of Alizarin Red S.

The materials used in the electrodes play a crucial role in determining how efficiently and environmentally friendly the wastewater is degraded during electrochemical wastewater treatment. The degradation of Alizarin Red was achieved in this bulk electrolysis experiment utilizing lead dioxide (PbO_2), BDD, and platinum (Pt) anodes in neutral media under the same circumstances. Using a BDD anode and a current density of 33 mA/cm^2, the TOC removal of the dye was accomplished promptly. PbO_2 and BDD anodes successfully removed the entire TOC and colour, whereas Pt anode only partially oxidized the dye. Kinetic investigations also revealed that PbO_2 and BDD electrodes had much greater oxidizing properties on dye than Pt anode [27].

Orange G

A Pt electrode is used to oxidize the azo dye Orange G (Fig. **8**), which is subject to the effects of the following variables: pH, current density, dye concentration (50–150 mg/L), rate of discoloration, rate of mineralization, current efficiency, and energy consumption. After 15 minutes of degradation in high chloride ion concentrations, all the distinctive UV-visible peaks of the chosen dye vanish. For wavelengths of 245 nm, 330 nm, and 480 nm, respectively, K_2SO_4 affects the elimination by 46%, 54%, and 61%. It suggests that the degradation processes in the environments of K_2SO_4 and KCl are distinct, and, that the elimination percentage of OG was unaffected by the rise in chloride concentrations. These findings support the idea that the anodic oxidation process is very beneficial for the discharge of concentrated contaminants [28].

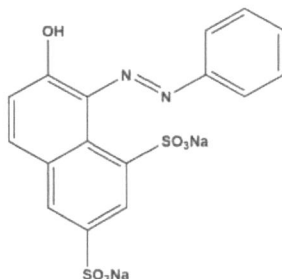

Fig. (8). Structure of Orange G.

Here, Orange G azo dye solutions have been electrochemically oxidized for discolouration and mineralization of 100 cm³ of 0.52-6.34 mmol/dm⁻³, pH 3.0, and current density between 33.3 and 150 mA/cm² in a divided or undivided tank reactor with the electrode combinations BDD/Stainless steel. The organic materials in both reactors were destroyed by hydroxyl radicals produced at the BDD anode by water oxidation. In a divided cell, complete decolorization and mineralization of the dye were successfully accomplished with a greater current density and shorter running period. It is concluded that a divided cell functions better than an undivided one. Total mineralization at 100 mA/cm² was likewise easily accomplished in the divided reactor at dye concentration 6.34 mmol/dm⁻³. Additionally, the entire mineralization efficiency was consistently obtained at higher dye concentration and lower current density. The rate constant of dye degradation improved obviously at larger current densities and followed a pseudo-first-order kinetics. Orange G was completely decoloured and eliminated in the same amount of electrolysis time, and as a result, the degradation byproducts were quickly converted into carboxylic acids like maleic, tartronic, acetic, formic, and oxalic acids and did not build up in the medium. Ammonium and sulfate were the primary inorganic ions generated throughout the mineralization process [29].

Acid-azo Dye

Acid Brown 14

A batch type electrochemical cell on a laboratory size was used to illustrate the acid brown 14 (AB-14) (Fig. **9**). The effects of current density, initial dye concentration, electrolyte (NaCl) concentration, pH on the efficacy of colour removal and COD were studied in two electrolysis procedures, namely anodic oxidation (AO) and electrocoagulation (EC). With increased current density and concentration, the efficacy of AB-14 dye degradation improves; however, degradation is less effective when pH and electrolyte concentration are increased. Additionally, it implies that under ideal and consistent experimental conditions, EC is more effective than AO. The kinetic model demonstrates that, depending on different operational factors, the electrolytic process follows the pseudo-firs--order-rate constant (k_{app}). The great efficacy of AO and EC treatments confirms that they are promising alternatives to traditional treatment techniques for the large-scale wastewater generated by the textile industry [30].

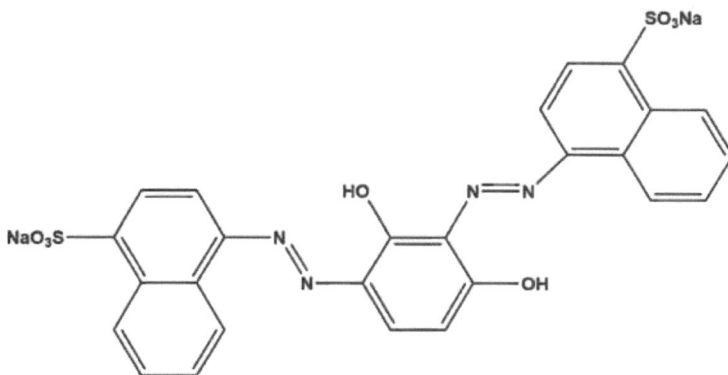

Fig. (9). Molecular structure of Acid Brown 14 dye.

Acid Orange 10

Two modified electrodes, Ti/PbO$_2$ and Ti/SnO$_2$-Sb, were used to electrochemically process the azo dye Acid orange 10 (Fig. **10**). The first anode was created using the electrochemical deposition technique, whereas the second was created using the traditional thermal breakdown technique. Current density, pH, and the concentration of the supporting electrolyte were used to evaluate the decolorization and degradation capacity of the dye on the surface of both anodes. The COD and TOC removal were calculated in order to evaluate the mineralization of AO10 in ideal circumstances. The ideal circumstances for the Ti/PbO$_2$ anode used in dye decolorization were a current density of 73.64 mA/cm^2, a pH of 12.05, and an electrolyte concentration of 117.04 mM. Due to these circumstances, decolorization was complete after 100 minutes of electrolysis, with 63% of COD and 60% of TOC removal being reported [31].

Additionally, the ideal parameters for the aqueous synthetic dye solution including Ti/SnO$_2$-Sb were mA/cm^2 current density, pH 2, and 75 mM/L electrolyte concentration. The prediction of the model was therefore tested, and the results showed that the minimum dye decolorization was 39% (current density = 6 and pH = 7.5), and the maximum dye decolorization was 101% (current density = 65 and pH = 2). Under the circumstances, 100 mg/L of dye was successfully decoloured, with 50 minutes of electrolysis resulting in COD and TOC removal rates of 61.3% and 43.9%, respectively. The Ti/SnO$_2$-Sb anode is superior to the other species in this study's analysis of the electrochemical degradation of AO10. The current density variable was used to determine decolorization. In conclusion, the electrochemical degradation of refractory dyes in aqueous solution using a Ti/SnO$_2$-Sb electrode is a suitable anode and an environmentally safe technique [32].

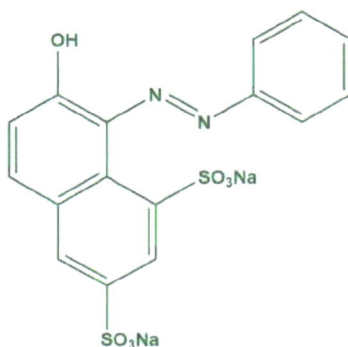

Fig. (10). Chemical structure of the azo dye Acid Orange 10.

Acid Red 1

At constant current density and pH 3.0, the azo dye Acid Red 1 (AR1) (Fig. **11**) underwent anodic oxidation in the presence of electrochemically produced H$_2$O$_2$ (AO-H$_2$O$_2$), electro-Fenton (EF), and photo-electro-Fenton (PEF). A Pt (or) BDD/air-diffusion electrode combination was used in this method to generate H$_2$O$_2$ from the reduction of O$_2$. The water source anodically produced hydroxyl radicals served as the main catalyst for the Fenton reaction between the produced H$_2$O$_2$ and 0.5 mmol/dm^3 Fe^{2+} in EF and PEF. The order of the electrode's oxidation characteristics was discovered to be AO-H$_2$O$_2$, EF and PEF. When compared to Pt, the BDD anode displayed superior reactivity. In the EF and PEF processes, the decolorization and obliteration of aromatics using hydroxyl radicals produced in large quantities was accomplished swiftly. Due to the quick photolysis of resistant intermediaries such Fe(III)-carboxylate complexes, the UVA irradiation PEF technique with BDD was the successful method in practically 100% mineralization. With an increase in current density, the electrogenerated hydroxyl radicals have a significant impact on the processes of mineralization and decolorization, but they have a negative impact on mineralization current efficiency. Totally, 11 aromatic intermediates, 15 hydroxylated compounds, 13 desulphonated derivatives, 7 short-linear carboxylic acids, and NH_4^{4+}, NO_3, and SO_4^{2-} ions were found during the azo dye degradation process. A comprehensive reaction sequence for AR1 mineralization is proposed using the electrolysis-produced chemicals [33].

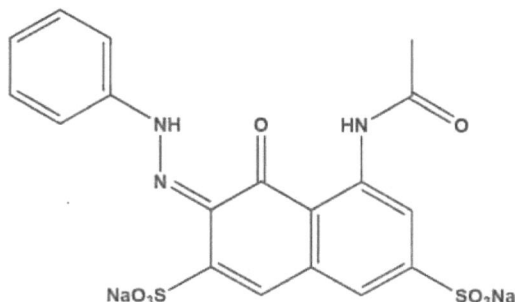

Fig. (11). Chemical structures of Acid Red 1.

Acid Red 18

Using sacrificial aluminium electrodes, a lab-scale EC batch reactor was used to remove the azo dye acid red 18 (Fig. **12**). The effects of several factors, including the beginning pH, current density, reaction time, initial dye concentration, electrode spacing, and type of electrolyte solution, on the efficiency of the EC process have been comprehensively investigated. At acidic pH 4, the ideal conditions were reached. The amount of dye removed grew along with the current density until it reached its maximum at 26 mA/cm². At current densities of 26 mA/cm² and 42 mA/cm², the anode mass reduction was increased from 202 mg and 331 mg, respectively. In addition, the mass of created sludge buildup increased from 605.85 mg to 1,060 mg, with a net mass of 400 mg. When electrodes were placed closer together, the decolorization was generally greater. The kind of electrolyte solution used had a significant impact on how much was consumed or how much energy was needed for decolorization [34].

Acid Blue 113

Galvanostatic electrolysis was used to successfully degrade Acid Blue 113 (Fig. **13**) in aqueous media. Here, lead dioxide, platinum supported on titanium, and TiO_2-nanotubes adorned with PbO_2 supported on titanium anodes were used to treat the electrochemical flow cell containing 250 mg/dm³ in 1.0 L industrial textile dye. Additionally, Na_2SO_4 was used as a supporting electrolyte at current densities of 20, 40, and 60 mA/cm². UV-visible spectroscopy and COD were used to periodically check the deterioration process. The outcomes showed that the dye was efficiently broken down on the Ti/TiO$_2$-nanotubes/PbO$_2$ electrode surface by electrochemically *in-situ* produced •OH from water discharge. In comparison to galvanostatic electrolysis utilizing Ti/Pt and Pb/PbO$_2$, the Ti/TiO$_2$-nanotubes/PbO$_2$ anode demonstrated an exceptional ability to remove the dye with better oxidation rate, higher current efficiency, and reduced energy consumption [35].

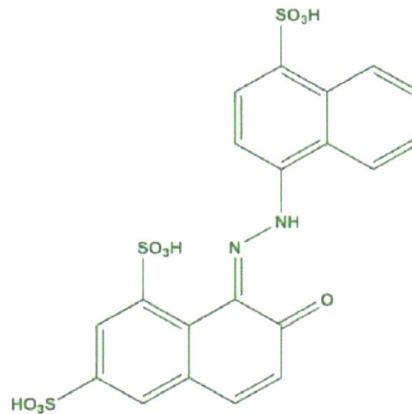

Fig. (12). Chemical structures of Acid red 18.

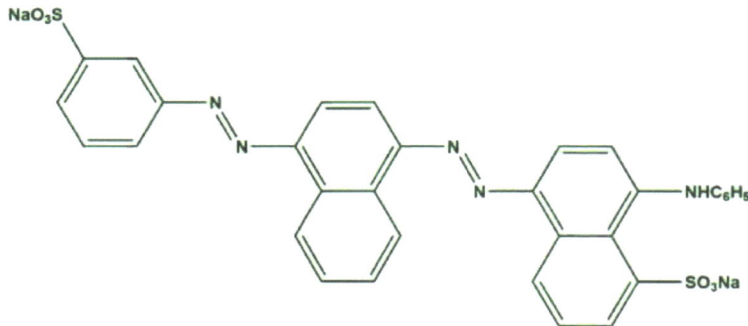

Fig. (13). Molecular structure of Acid Blue 113.

Methylene Blue and Methyl Blue

In a brand-new electrochemical setup with chloride ions, a unique method of anodic two-dye degradation, including methylene blue (Fig. **14a**) and methyl blue (Fig. **14b**), was carried out. The cell is composed of sandwich-style electrode configurations, where a group of graphite rods acting as the anode are positioned horizontally between two horizontal stainless-steel screen cathodes. Along with electrical energy consumption, COD reduction, anode efficiency, and current efficiency, the performance of the cell is also tracked. Some of the variables that were looked at included the starting pH of the solution, initial dye concentration, initial NaCl concentration, and applied current density. Methylene blue and methyl blue had maximum colour removal rates of 94% and 98%, respectively, after 30 minutes of treatment. The standardized space velocity of methylene blue in terms of kinetic parameter value was determined to be 3.59 m^3h^{-1} and 4.09 m^3h^{-1} for methyl blue itself [36].

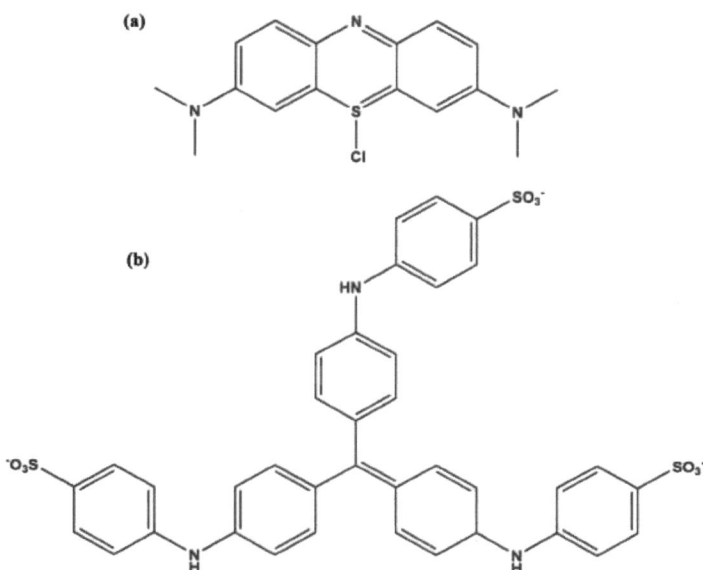

Fig. (14). Chemical structures of the model dyes: (**a**) methylene blue, (**b**) methyl blue.

A creative alternate method for the electrochemical methods for treating wastewater including colours was also developed by Oliveira, G.R. and colleagues. This study investigated anodic oxidation employing Ti/Pt anodes as a treatment option for synthetic wastewaters that contained Methylene Blue as a model dye. At various operating circumstances, such as current density, agitation rate, and temperature, galvanostatic electrolysis of MB synthetic wastewaters has resulted in the total elimination of discolouration. The impact of key operational parameters (current density, agitation rate, and temperature) was examined in order to determine the ideal electrolysis conditions for TOC and COD reduction. Based on the electrocatalytic qualities of the Ti/Pt anode and experimental findings, it can be concluded that the electrochemical oxidation process effectively reduces COD and discolours wastewater containing MB dye. The amount of energy required to remove colour from MB synthetic solutions during galvanostatic electrolysis mainly depends on the applied current density, temperature, and agitation rate; for instance, it ranges from 7.95 kWh at 10 mA/cm² to 33.60 kWh at 50 mA/cm² per volume of treated effluent removed (m³); from 18.78 kWh/m³ at 25 °C to 10.29 kWh/m³ at 60 °C and from 21. The electrochemical technique may be a workable substitute for decolorizing wastewaters that contain dyes, considering the results obtained [37].

Methyl Orange

A comparative electrochemical oxidation of the azo dye methyl orange (Fig. **15**) was described by E. Isarain-Chávez *et al.* utilizing a variety of electrodes as anodes, including Ti/Ir-Pb, Ti/Ir-Sn, Ti/Ru-Pb, Ti/Pt-Pd, and Ti/RuO₂. In this study, the anodic oxidation of 2 dm³ of methyl orange azo dye solutions in 0.050 mol/dm⁻³ Na₂SO₄ of pH 7.0 at constant current density was used to determine the relative oxidation potential of dimensionally stable anodes (DSA). This process revealed the compositions of Ti/Ir-Pb, Ti/Ir-Sn, Ti/Ru-Pb, Ti/Pt-Pd, and Ti/RuO₂. Using the dip-coating method, the anodes were made, and their composition and morphology were examined. The Ti/Ir-Pb anode that was constructed of a combination of IrO₂, Pb₂O₃, and Pb₃O₄ worked better. The effects of current density, Na₂SO₄ concentration, and cathode type on the anodic oxidation with Ti/RuO₂ decolorization of azo dye solutions were investigated. It is interesting to note that 96%–98% colour removal was achieved utilizing Ti/Ir-Pb, Ti/Ir-Sn, and Ti/Ru-Pb anodes, but Ti/Pt-Pd and Ti/RuO₂ anodes decoloured less under the ideal circumstances. The decolorization process was discovered to follow pseudo-first-order kinetics in each of these scenarios. The oxidation capacity of the anodes improved in the following order: Ti/RuO₂; Ti/Pt-Pd; Ti/Ru-Pb; Ti/Ir-Sn; and Ti/Ir-Pb, with the latter electrode achieving 76.0% mineralization. Because it increased the oxidation efficiency of hydroxyl radicals produced in the non-active oxide, the combination of active and non-active materials produced anodes with a higher oxidation potential than those made completely of active materials.

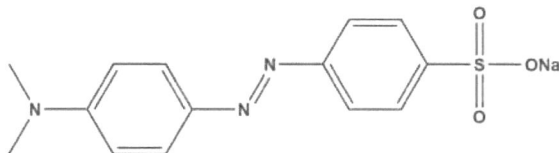

Fig. (15). Molecular structure of methyl orange azo.

Ir has an advantage over Ru in mixed metal oxides because organic molecules adhere to its surface better, promoting their oxidation. Sulphate and ammonium ions were released as the dominant ions. The stable byproducts and final short-linear aliphatic carboxylic acids were identified using ion-exclusion high-performance liquid chromatography (HPLC) and gas chromatography-mass spectrometry (GC-MS). Based on these analytical characterizations, the mineralization pathways and the mechanism of methyl orange dye degradation transition were assessed [38].

Solutions of the azo dye methyl orange were destroyed by electrochemical oxidation in Ramirez C and coworkers' investigation using a 3 L flow plant with a boron-doped diamond (BDD)/stainless steel cell operating at constant current density, ambient temperature, and a liquid flow rate of 12 L min-1. A 23-factorial design with the applied current density, azo dye concentration, and electrolysis time as variable independents was used to evaluate the technique using the response surface approach. The -N=N group of the dye, according to LC-MS analysis, was broken down into around seven oxidation products, which were then deaminated, formed into a nitro group, and/or desulphonated to produce aromatics [39].

Turquoise Blue GB

M. Wen *et al*. [40] investigated the electrochemical oxidation-induced breakdown of cationic turquoise blue GB using distinctive rare earth doped PbO_2 electrodes. By using a thermal decomposition-electrodeposition approach, the unique PbO_2 doped with typical rare earth (Ce) electrode was created in this study. The impacts of electrode materials used as anodes in electrochemical oxidation systems were carefully examined for GB degradation. A comparison was made between the effects of Ce-doped and undoped electrodes on dye deterioration. The ideal degradation conditions were investigated with Ce-doped electrode based on the effects of various parameters, such as current densities and electrolyte concentrations. The properties of electrodes that were both undoped and cemented were examined using SEM images.

Remazol Brilliant Blue R

By using the electrodeposition procedure with an alkaline solution, Mukimin and colleagues created a modified electrode of titanium-lead dioxide (Ti/PbO_2) that contained nano-rod-shaped PbO_2 particles [41]. The electrode exhibits electroactive characteristics, which are demonstrated by the cyclic voltammetry (CV) diagram. In order to electro-degrade Remazol Brilliant Blue R, the electrode was employed as the anode (Fig. **16**). In order to quantify the effects of electro-degradation, UV-Vis, COD, and HPLC studies were used. In UV-Vis examination, a fast decreasing absorbance was seen in the visible area (592 nm), which suggests that the degradation is easily started by the breakage of the link between the anthraquinone and 1,4-diNHArt group. The absorbance at 227 nm and 286 nm, however, showed that the degradation occurred over a longer time when the link on the anthraquinone molecule was broken. The dye had been broken down into simple chemicals or small organic molecules, according to

HPLC analyses. Furthermore, the fraction of COD is decreased to 70.38%, indicating the significant potential and possibility for implementing an electrolysis approach in the future utilizing this kind of electrode. With a pH range of 5 to 10, 4000 mg/L of NaCl, and a degradation time of 50 to 60 minutes, electro-degradation is effective.

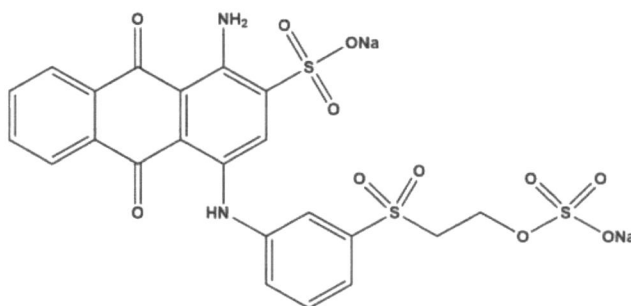

Fig. (16). Molecular structure of Remazol Brilliant Blue R.

Novacron Yellow (NY) and Remazol Red (RR)

By using Pt supported on Ti as the anode, C. K. C. Arajo *et al.* demonstrated the efficiency of electrochemical oxidation (EO) for the removal of a dye mixture combining Remazol Red (RR) and Novacron Yellow (NY) (Fig. **17**) in aqueous solutions. Various current densities, including 20, 40, and 60 mA/cm^2, and temperatures, including 25, 40, and 60 °C, were tested in electrochemical treatment to improve the reaction conditions. The EO of each of these dyes was subsequently investigated independently. Each of these dyes, EO was conducted at different current densities while maintaining the same temperature (25 °C). By using UV-visible spectroscopy, the decolorization was noticed, and COD was used to examine the decomposition of organic components.

According to the data obtained from the examination of the dye mixture, the EO approach was effective in eliminating colour because more than 90% of it was eliminated. In case of COD elimination, no complete oxidation was achieved since the oxygen evolution reaction was promoted by a current density larger than 40 mA/cm^2. In regard to the analysis of individual anodic oxidation dyes, it was noted that the data for the NY dye were quite similar to the outcomes of the oxidation of the dye mixture, whereas the RR dye had a greater capacity to remove colour but less capacity to remove COD. These findings show that the type of organic molecule affects the oxidation efficiency, and this was supported by the discovered intermediates [42].

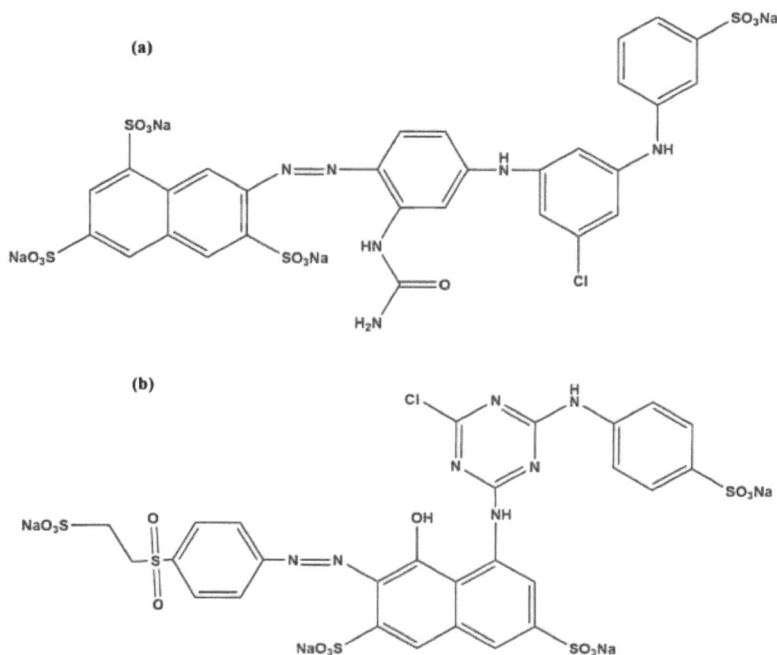

Fig. (17). Chemical structures of (**a**) NY and (**b**) RR.

Eosin Y and Rose Bengal

Tabarra *et al.* used boron doped diamond (BDD) and Pt electrodes to investigate the indirect anodic oxidation-induced decolorization of Eosin Y (Fig. **18**) and Rose Bengal (Fig. **19**). The influence of electrolyte, starting pH, and applied current density on the mineralization behaviour of xanthene dyes was investigated. With the use of a UV-Vis spectrophotometer, the decolorization and degradation of dyes were periodically monitored. *In situ*, production of the •OH radical, $S_2O_8^{2-}$ at the BDD electrode, and active chlorine species (Cl_2, HOCl) at the Pt electrode were used to explain the results. The production of refractory chlorinated organic compounds was thought to be the cause of poor mineralization at both BDD and Pt anodes when chloride was present. Effective mineralization of xanthene dyes is accomplished at the BDD electrode while employing Na_2SO_4 as the supporting electrolyte [43].

Rhodamine B

Arajo *et al.* seek to use Conductive Diamond Electrochemical Oxidation (CDEO) to have mediated oxidation affect the removal of Rhodamine B (Fig. **20**) solutions. Four different supporting electrolytes, including Na_2SO_4, $HClO_4$,

H_3PO_4, and NaCl, were employed to achieve this electrochemical reaction. Although the supporting electrolyte medium significantly affects the rate of operations and efficiency, CDEO entirely eliminates organic pollutants regardless of the medium. Accordingly, it is stated that electrolysis in sulphate and phosphate media act similarly, although it behaves remarkably better in perchlorate media than in chloride media. In general, the current density is very important. The CDEO follows a first-order kinetic in all optimization experiments, and the kinetic constants are typically substantially higher than would be predicted from a single mass transfer electrolytic model. This shows the value of using mediated electrochemical methods to remove Rhodamine B because it is not what a direct electrochemical oxidation procedure should have generated. The oxidation conditions of CDEO yield fewer reactive intermediates than those of other methods. The intermediates (aromatic acids) generated in the early stages of the process are quickly mineralized to CO_2 when the short-chain aliphatic acids that were present are eliminated. In chloride media, hypochlorite leads to the production of chlorinated intermediates [44].

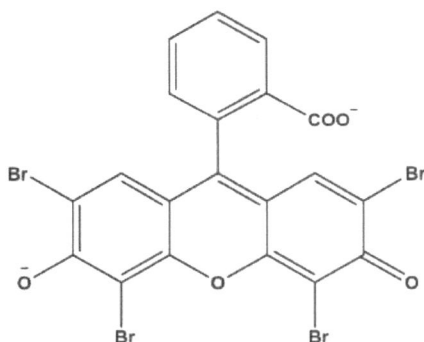

Fig. (18). Chemical structure of Eosin Y.

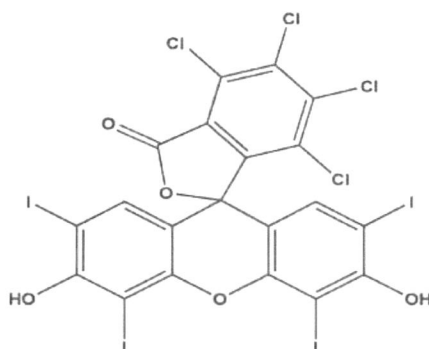

Fig. (19). Chemical structure of Rose Bengal.

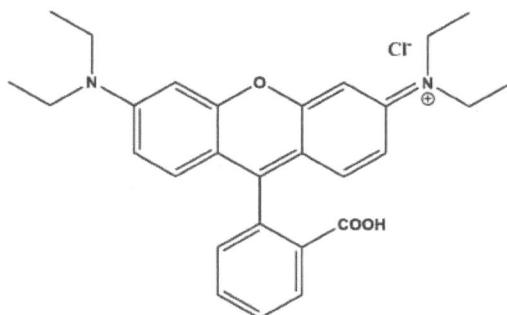

Fig. (20). Chemical structures of Rhodamine B.

The electrochemical oxidation efficiency of Ti/RuO$_2$-IrO$_2$ (DSA) and SnO$_2$ anodes on the oxidation of Rhodamine B (RhB) dye was also examined by Ali Baddouh and colleagues. The optimization studies of current density, initial solution pH (pH0), type and concentration of electrolyte, and temperature during the electrochemical oxidation were studied in order to evaluate the decolorization and COD elimination process. Using both DSA and SnO$_2$ electrodes, complete decolorization of RhB was accomplished in the presence of chloride ions. The DSA electrode was used to obtain the highest efficiency at a pH of 6.5, a temperature of 25 °C, a current density of 40 mA/cm^2, and a mixture of NaCl and Na$_2$SO$_4$ as the supporting electrolyte. After 90 minutes of electrolysis under this electrochemical oxidation setting, 100% colour removal and 61.7% chemical oxygen demand elimination were accomplished. Under demanding working conditions, DSA outperformed SnO$_2$ and proved to be more economical and efficient. The efficiency of the degradation is explained by indirect electrochemical oxidation, in which the presence of chlorides causes the electrolyte to produce potent oxidizing species such as Cl$_2$ and ClO- ions, which enhance the efficacy of treatment at both electrodes [45].

Auramine-O

Auramine-O was electrochemically oxidized by Hmani *et al.* utilizing galvanostatic electrolysis with boron-doped diamond (BDD) and lead dioxide (PbO$_2$) anodes (Fig. **21**). The UV-visible spectrometry and the COD were used to track the electrolysis's progress. The outcomes showed that both electrodes successfully decomposed auramine-O (AO). Through electrogenerated hydroxyl radicals from water discharge on the electrode surface, auramine-O is electrochemically degraded. Auramine-O's rate of deterioration is highly influenced by its initial concentration, current density, and pH. On the other hand, the process of degradation is not much impacted by temperature. Additionally, the

COD decay has a pseudo-first-order kinetic and was governed by mass transfer. The outcomes demonstrated that under ideal experimental circumstances, such as COD_0 150 mg/L, japp50 mA/cm^2, pH 6, and 30 °C, approximately 95% of COD removal is accomplished. Among these electrodes, the BDD anode outperformed PbO_2 in its ability to remove AO. It offered a greater rate of oxidation, and a higher current efficiency, and used less energy than galvanostatic electrolysis using the PbO_2 electrode [46].

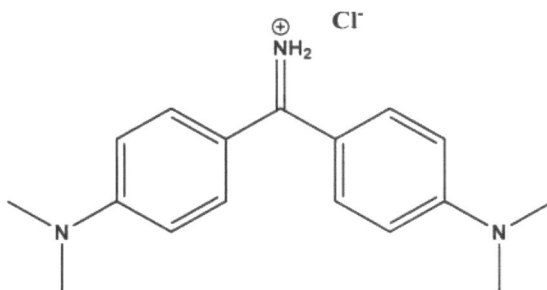

Fig. (21). Molecular structure of auramine-O (AO).

Allura Red AC

With electrogenerated H_2O_2 (EO-H_2O_2), electro-Fenton (EF), and photoelectro-Fenton (PEF), Thiam, A. *et al.* devised an electrochemical approach for the decolorization and mineralization of solutions containing 230 mgL^{-1} of the food azo dye Allura Red AC (Fig. **22**) at pH 3.0. A stirred tank reactor with a boron-doped diamond (BDD) or Pt anode and an air-diffusion cathode was used in trials to create H_2O_2. Hydroxyl radicals are produced by the Fenton reaction between the bulk H_2O_2 and Fe^{2+} as well as by the water oxidation at the anode surface. The primary oxidants were these hydroxyl radicals. The ability to oxidize was strengthened by the sequence EO-H_2O_2 EF-PEF, and BDD was always utilized to quicken degradation. In SO_4^{2-}, ClO_4^-, and NO_3^- media, the PEF process with BDD exhibited nearly entire mineralization following similar trends, however in the Cl$^-$ medium, mineralization was hampered by the generation of refractory chlorine derivatives. Gas chromatography-mass spectrometry (GC-MS) research proved that the -N=N bond was broken, resulting in the creation of two major aromatics in SO_4^{2-} medium and three chloro aromatics in Cl$^-$ solutions. Due to BDD's effective oxidation of the final oxalic and oxamic acids as well as UVA light's photolysis of Fe(III)-oxalate species, PEF with BDD has an advantage over other methods. NH^{4+}, NO_3^-, and SO_4^{2-} ions were released as a result of mineralization [47].

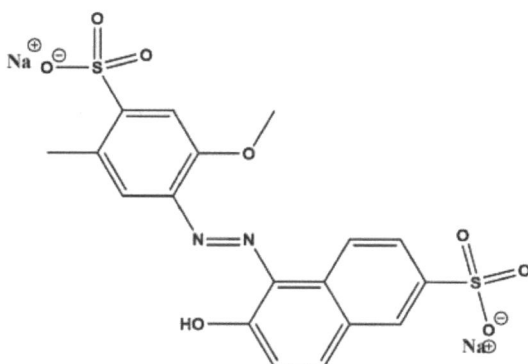

Fig. (22). Molecular structure of Allura Red AC.

Direct Red 81

The electrocoagulation method for the decolorization of solutions containing Direct red 81 was described by S. Aoudj *et al.* (Fig. **23**). Synthetic solution experiments were run in the batch mode. The effects of key variables, including electrolysis time, current density, initial pH, inter-electrode distance, initial dye concentration, and type of supporting electrolyte, were examined in order to improve the reaction conditions. Thus, starting pH of around 6, current density of 1.875mA/cm^2, inter-electrode distance of 1.5 cm, and lastly the usage of table salt NaCl as supporting electrolytes are the ideal reaction conditions for decolorization. Additionally, this ideal circumstance offers a high degree of decolorization efficiency (more than 98% of colour removal was accomplished). The remaining EC byproduct was evaluated using Fourier transform-infrared spectroscopy (FT-IR) analysis with and without dye [48].

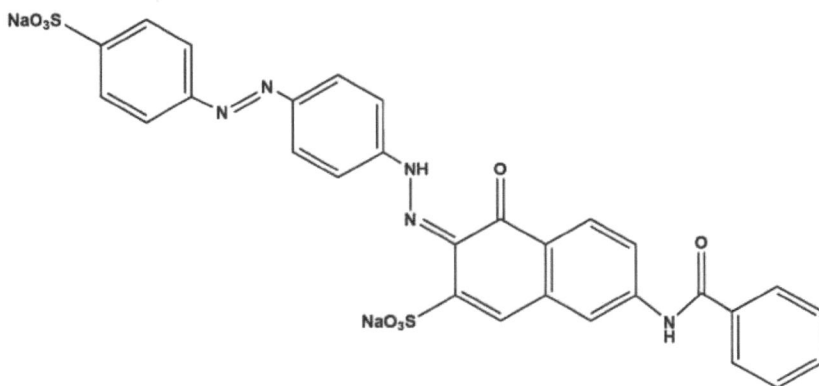

Fig. (23). Molecular structure of Direct red 81.

Congo Red

A method using a modified anode containing polyaniline (PANI) and graphene that was created by the electro-deposition process was proposed by Ruixiang Li *et al*. As a result, the greatest degree of decolorization effectiveness was achieved. It's interesting to note that at 54 hours, the Congo red (CR) (Fig. **24**) decolorization rate in the microbial electrochemical systems (MESs) with the PANI/graphene-modified electrode (PG) achieved nearly 90%. But in MESs with a PANI-modified electrode (P) and MESs with an unmodified electrode (C), the rates of CR decolorization were only 68% and 79%, respectively (Scheme 3). The abundant *Methanobrevibacterarboriphilus* in PG (11%), which was 5.5 times that in C (2%) at 18 h, was highlighted by the results of the microbial community study. It is possible to link this occurrence to the quick decolorization. The upregulated metabolic pathways, such as those for arginine and proline metabolism, purine metabolism, arginine biosynthesis, and riboflavin metabolism, improved extracellular electron transport by increasing the availability of electron shuttles and redox mediators. The PG-modified electrode promoted the decolorization as a result by altering particular metabolic pathways. The guidelines for the potential use of MESs for wastewater treatment may be improved by this research [49].

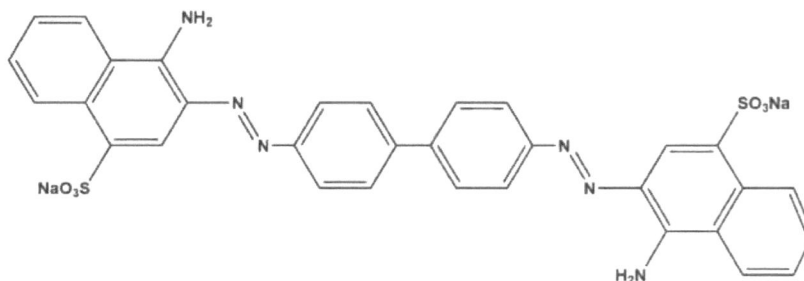

Fig. (24). Chemical structures of Congo red.

Amido Black 10B (AB)

To examine the sedimentation kinetics of AB, a simple electrocoagulation induced settling tank reactor (EISTR) was used and optimized (Fig. **25**). Additionally, tertiary and binary dye molecular combinations were used to explain the effectiveness of the dye removal by EISTR [50]. To increase AB dye removal efficiency, the most crucial operating parameters, including current density, electrolysis time, initial solution pH, and electrolyte concentration, were improved. The chemical oxygen demand (COD), chemical reaction efficiency (CRE), starting sludge settling velocity, final sludge height, and turbidity decrease were all measured. These investigations demonstrate that 81-99% CRE, 54-68%

COD reduction, and 73-88% turbidity reduction are possible in synthetic and textile dye wastewater systems. It was found that for the majority of the dyes, 3.8 Ah of charge loading per gram of dye was sufficient to provide high CRE and fantastic sedimentation kinetics. This work showed that, under similar experimental conditions, Fe electrodes outperform Al electrodes. In the end, the study suggested that the removal of dye molecules, dye molecular combinations, and textile effluent can all be accomplished with a single EISTR utilizing Fe-electrodes.

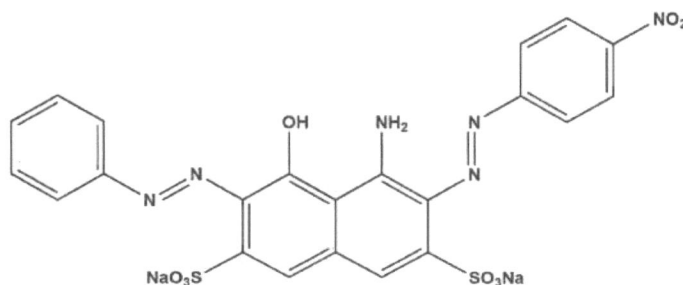

Fig. (25). Chemical structures of Amido Black 10B dye.

The same authors also looked at the role that common oxidants including peroxomonosulfate (PMS), peroxodisulfate (PDS), and hydrogen peroxide have in helping the AB dye break down during electrocoagulation (EC) processes (HP) [51]. In order to create hydroxyl (HO•) and sulphate (SO_4•$^-$) radicals, these EC activities were dependent on the electrochemically generated Fe^{2+}/Fe^{3+}-mediated activation of PMS, HP, and PDS. It took a lot of work to figure out how the starting solution's pH, the applied current, and the concentration of the different oxidants affected the way AB dye deteriorated. Experimentation with Fe electrodes confirmed the ideal operating conditions for each of these features.

This study showed that oxidant-assisted EC processes performed best and were most effective under low applied current and an acidic environment (pH 5). The experimental findings showed that, at the optimum concentrations of CAB = 0.16 mM, CNaCl = 17 mM, CPMS = CPDS = 0.16 mM, CHP = 0.13 mM, and pH 5, oxidant-assisted EC processes produced impressive AB dye degradation (99%) with little energy expenditure. The order in which amido black 10B dye was most efficiently destroyed by oxidant-assisted EC was PMS > PDS > HP. The azo linkage (-N=N-) of the AB dye molecule was degraded by oxidant-assisted EC processes more quickly than the aromatic rings, according to the UV-visible spectral alterations. The highest mineralization efficiencies were 34, 66, 57, and 46% for EC and EC supplemented by PMS, PDS, and HP, respectively.

Effect of Microbial Fuel Cell (MFC) on Dye Degradation

Microbial fuel cells (MFCs) were employed by Carmalin, S.A. *et al.* to treat various kinds of organic effluents by using bacteria as biocatalysts and generating electrical energy [52]. The electrodes are essential to the MFC bioelectrochemical reactions. Battery carbon rods were used as the cathode material in the current experiment to break down and discolour textile dyes. Utilizing carbon rods from used zinc-carbon cells is a novel and environmentally friendly method of recycling trash in an MFC. Real textile dye effluent had been put onto the MFC anode. The MFC under study achieved a maximum output of 56.42 mW over a period of 600 hours, while also lowering COD by 90%. A 28.5% decolorization level was attained. According to the field emission-SEM study of the biofilm, it was composed of thick, opaque sheets that were likely exopolysaccharide-coated. Under optical microscopy, the biofilm's bacteria seemed to be isolated and bilobed. The carbon rods of zinc carbon cells can be utilized as an MFC cathode to simultaneously generate power and degrade dye, according to the findings of a study.

Although adding a redox mediator (RM) to the MFC azo dye decolorization process can successfully improve the electrogenesis and decolorization performance, RMs are wasted when batch runs use different substrates. In order to address this problem, Huang, W. and coworkers created RM-modified anodes by electrodepositing riboflavin (RF) and humic acid (HA) on the surface of graphite felt. These anodes were then used to build air-cathode single-chamber MFCs with various modified anodes in order to remove the colour of Congo red and generate electricity. When compared to bare anode MFC, modified anodes with 0.5C RF, 0.5C HA, 1.25C RF, and 2.5C HA showed good electrocatalytic activity, 31%, 34%, 44%, and 49% decreases in internal resistance, and 20%, 21%, 40%, and 66% improvements in maximum power density. The high Congo red decolorization efficiencies of MFCs with 2.5C HA and 1.25C RF modified anodes are 86% and 75% in 16 hours, respectively. The modified anodes had thick RM crystals and bacterial colonies on their surfaces after MFC was operated. By accelerating electron transmission, the RM crystals on the modified anodes aid in decolorization and the generation of bioelectricity [53].

Electrode Materials-effect of Metal Oxide Electrodes in the Degradation Process

In general, several metal oxide electrode materials function as significant critical parameters in the demineralization and removal of colour from dyes. According to Comninellis *et al.* [54], the type of electrode material has a considerable impact on the process' selectivity and efficiency. Different anodes support partial and selective oxidation of pollutants while others favour the entire conversion to CO_2. They provided a thorough model for the oxidation of organic materials at metal oxide electrodes with concomitant oxygen evolution in order to interpret them. They presented a theory that relies on the distinction between "active" and "non-active" anodes to explain how hydroxyl radicals can oxidize organic molecules. The "active" anodes include Pt, IrO_2, and RuO_2. On the other side, "non-active" anodes include PbO_2, SnO_2, and boron-doped diamond (BDD). An electrode is said to be "non-active" if it does not offer an active catalytic site for the adsorption of materials from the aqueous environment. Here, the anode only acts as a well for the extraction of electrons and an inert substrate [55].

RuO₂

P. Kaur and colleagues examined the effects of electro-oxidation parameters such as pH, current, and electrolysis time on percentage COD removal, percentage colour removal, and energy consumed. They used a Ti electrode coated with RuO_2 (Ti/RuO_2). The Box Behnken Design was applied to both the data analysis and the experimental planning. In addition, studies on the direct and/or indirect oxidation of contaminants were conducted. GC-MS and spectrophotometric analysis were utilized to identify the eliminated organic components and transformation products in treated effluent, as well as to determine the safe disposalability of treated wastewater. To determine whether the procedure was commercially viable, an operating cost (electrode and electricity cost) study was also completed. Most of the organics were removed during the electro-oxidation process, according to a GC-MS study of the wastewater that had been electro-oxidation treated. Chlorinated organic molecules and other compounds that had been changed into other compounds were also present in untreated textile effluent [56].

TiO₂

The performance of the electro-oxidation procedure for discolouring the textile wastewater effluent pre-treated utilizing a lab-scale moving bed-membrane bioreactor was examined by Bakaraki Turan, N. *et al.* Graphite-TiO_2, TiO_2, TiO_2-coated Platine, and TiO_2-coated ruthenium dioxide (RuO_2) were tested and assessed for their colour removal efficiency as cathodic electrode materials. TiO_2 was employed as the anode electrode. Additionally, the investigation focused on the electrode material-related optimization parameters, including pH, electrolysis time, and applied current. The observed decolorization percentages for TiO_2/Graphite, TiO_2/Platine, TiO_2, and TiO_2/RuO_2 were 92.95%, 91.58%, 91.40%, and 89.17%, respectively. In this situation, the best values for each electrode material were chosen. Finally, in order to make it easier and more useful for readers to select and evaluate the electrode materials, the operational costs for each tested cathodic electrode material were computed in each of the explored optimization parameters. The correlation coefficients (R2) in the optimization study using nonlinear regression modelling were 81.2%, 87.1%, 86.7%, and 88.6%, respectively [57].

Boron-Doped Diamond (BDD)

Due to its broad potential window, low background current, very low activity for the O_2 evolution process, and high anodic potential, the BDD-based anode proposed by J. T. Matsushima *et al.* has garnered a lot of interest in the wastewater treatment industry [58]. The behaviour of the BDD/Ti and NCD/Ti electrodes was nearly reversible. The BDD/Ti electrode, on the other hand, displayed the fastest electron transport kinetics. The quasi-reversible behaviour of the investigated redox reactions may be attributed to the low electron transfer kinetics of these electrodes, which are insufficient to maintain the species' equilibrium in the solution interface. Tsantaki E *et al.* investigated the electrooxidation of textile effluents over a boron-doped diamond anode. Both an actual textile dyeing process effluent and a multi-component synthetic solution made up of 17 dyes and several supplementary inorganics were used in the experiments for this investigation. Following variations in total organic carbon, chemical oxygen demand, and colour, the effect of varying operating settings, such as current density (4-50mA/cm^2), electrolyte concentration (0.1-0.5M $HClO_4$), initial solution pH (1-12.3), and temperature (22-43 °C), on process efficiency was examined. Complete decolorization and substantial mineralization (up to 85% depending on the condition) might be accomplished after 180 minutes of treatment. Higher electrolyte concentrations and lower pH levels improved performance, although the temperature had little impact. Energy was used more efficiently per unit mass of COD removed at lower current densities than at higher densities, where energy was unnecessarily wasted on side reactions [59].

Platinum

The Pt electrode is one of the most often used anodes in both preparative electrolysis and synthesis due to its strong chemical resistance to corrosion even in very aggressive conditions. The behaviour of platinum electrodes in the electrochemical oxidation of organic contaminants has been well-researched in the literature and has been found to exhibit significant electrocatalytic activity [60].

Titanium–tantalum–platinum–iridium

E. Chatzisymeon *et al.* successfully investigated the electrochemical oxidation of textile effluents over a titanium-tantalum-platinum-iridium anode. Batch tests using highly coloured synthetic effluent containing 16 textile colours at a total concentration of 361 mg/L and a COD of 281 mg/L were carried out in a flow-through electrolytic cell with internal recirculation at current intensities of 5, 10, 14, and 20 A, respectively, and NaCl concentrations of 0.5, 1, and 4%. A dyehouse sewage sample with a COD of 404 mg/L was also examined. In addition to a range of other inorganic and organic substances, it contained leftover colours. Quantitative effluent decolorization was typically attained after 10 to 15 minutes of treatment, using little energy; in contrast, the extent of mineralization varied between 30 and 90% after 180 minutes, depending on the operating parameters and the type of effluent. In general, treatment effectiveness rose with rising salinity and current strength and fell with rising solution pH. Degradation was however minimized by utilizing electrolytes devoid of chloride, such as $FeSO_4$ or Na_2SO_4. Although the effluent's acute toxicity to the marine bacteria *Vibrio fischeri* was low, it considerably increased following treatment, which indicated the generation of harmful byproducts with a lengthy half-life [61].

CONCLUSION

Electrochemical anodic dye degradation work has advanced to a promising stage of development and is now capable of effectively destroying toxic or biorefractory organic dyes. It is a much more pronounced environmentally benign process, offering resourcefulness, energy efficiency, amenability to automation, environmental compatibility, and cost-effectiveness. Many businesses use this approach for bulk degradation because of the simple operating manner and efficient electrochemical process performance. The current chapter discusses a broad range of azo dye degradation, the impact of various electrode materials, and degradation caused by electrocoagulation techniques. Additionally, the mineralization and degradation processes were carried out at ideal electrolysis conditions, including pH, current density, and supporting electrolyte content, and they were assessed using a variety of analytical techniques, including UV-vis, TOC, and COD analysis.

REFERENCES

[1]   J.J. Roxon, A.J. Ryan, and S.E. Wright, "Reduction of water-soluble azo dyes by intestinal bacteria", *Food Cosmet. Toxicol.,* vol. 5, no. 3, pp. 367-369, 1967.
        [http://dx.doi.org/10.1016/S0015-6264(67)83064-5] [PMID: 6054354]

[2]   D. Brown, "Effects of colorants in the aquatic environment", *Ecotoxicol. Environ. Saf.,* vol. 13, no. 2, pp. 139-147, 1987.
        [http://dx.doi.org/10.1016/0147-6513(87)90001-7] [PMID: 3595482]

[3]   M. Qamar, M. Saquib, and M. Muneer, "Semiconductor-mediated photocatalytic degradation of anazo dye, chrysoidine Y in aqueous suspensions", *Desalination,* vol. 171, no. 2, pp. 185-193, 2005.
        [http://dx.doi.org/10.1016/j.desal.2004.04.005]

[4]   S. El Aggadi, N. Loudiyi, A. Chadil, O. Cherkaoui, and A. El Hourch, "Electrochemical oxidation of

textile azo dye reactive orange 16 on the Platinum electrode", *Mediterr. J. Chem.,* vol. 10, no. 1, pp. 82-89, 2020.
[http://dx.doi.org/10.13171/mjc10102001311108sea]

[5]   K.P. Sharma, S. Sharma, S. Sharma, P.K. Singh, S. Kumar, R. Grover, and P.K. Sharma, "A comparative study on characterization of textile wastewaters (untreated and treated) toxicity by chemical and biological tests", *Chemosphere,* vol. 69, no. 1, pp. 48-54, 2007.
[http://dx.doi.org/10.1016/j.chemosphere.2007.04.086] [PMID: 17583772]

[6]   X. Liu, J. Tian, Y. Li, N. Sun, S. Mi, Y. Xie, and Z. Chen, "Enhanced dyes adsorption from wastewater *via* Fe_3O_4 nanoparticles functionalized activated carbon", *J. Hazard. Mater.,* vol. 373, pp. 397-407, 2019.
[http://dx.doi.org/10.1016/j.jhazmat.2019.03.103] [PMID: 30933862]

[7]   E. Forgacs, T. Cserháti, and G. Oros, "Removal of synthetic dyes from wastewaters: A review", *Environ. Int.,* vol. 30, no. 7, pp. 953-971, 2004.
[http://dx.doi.org/10.1016/j.envint.2004.02.001] [PMID: 15196844]

[8]   M. Rajabi, K. Mahanpoor, and O. Moradi, "Removal of dye molecules from aqueous solution by carbon nanotubes and carbon nanotube functional groups: Critical review", *RSC Advances,* vol. 7, no. 74, pp. 47083-47090, 2017.
[http://dx.doi.org/10.1039/C7RA09377B]

[9]   D. Bhatia, N.R. Sharma, J. Singh, and R.S. Kanwar, "Biological methods for textile dye removal from wastewater: A review", *Crit. Rev. Environ. Sci. Technol.,* vol. 47, no. 19, pp. 1836-1876, 2017.
[http://dx.doi.org/10.1080/10643389.2017.1393263]

[10] V. Katheresan, J. Kansedo, and S.Y. Lau, "Efficiency of various recent wastewater dye removal methods: A review", *J. Environ. Chem. Eng.,* vol. 6, no. 4, pp. 4676-4697, 2018.
[http://dx.doi.org/10.1016/j.jece.2018.06.060]

[11] S. Song, J. Fan, Z. He, L. Zhan, Z. Liu, J. Chen, and X. Xu, "Electrochemical degradation of azo dye C.I. Reactive Red 195 by anodic oxidation on Ti/SnO_2–Sb/PbO_2 electrodes", *Electrochim. Acta,* vol. 55, no. 11, pp. 3606-3613, 2010.
[http://dx.doi.org/10.1016/j.electacta.2010.01.101]

[12] J. Sun, H. Lu, L. Du, H. Lin, and H. Li, "Anodic oxidation of anthraquinone dye Alizarin Red S at Ti/BDD electrodes", *Appl. Surf. Sci.,* vol. 257, no. 15, pp. 6667-6671, 2011.
[http://dx.doi.org/10.1016/j.apsusc.2011.02.099]

[13] W. Yun-Hai, C. Qing-Yun, L. Guo, and L. Xiang-Li, "Anodic materials with high energy efficiency for electrochemical oxidation of toxic organics in waste water", In: *in: Ind. Waste* InTech, 2012.
[http://dx.doi.org/10.5772/38556]

[14] L. Feng, E.D. van Hullebusch, M.A. Rodrigo, G. Esposito, and M.A. Oturan, "Removal of residual anti-inflammatory and analgesic pharmaceuticals from aqueous systems by electrochemical advanced oxidation processes. A review", *Chem. Eng. J.,* vol. 228, pp. 944-964, 2013.
[http://dx.doi.org/10.1016/j.cej.2013.05.061]

[15] M. Panizza, and G. Cerisola, "Direct and mediated anodic oxidation of organic pollutants", *Chem. Rev.,* vol. 109, no. 12, pp. 6541-6569, 2009.
[http://dx.doi.org/10.1021/cr9001319] [PMID: 19658401]

[16] M. Panizza, *Organic Pollutants, Direct and Mediated Anodic Oxidation.* Encycl. Appl. Electrochem., Springer New York, 2014, pp. 1424-1428.
[http://dx.doi.org/10.1007/978-1-4419-6996-5_126]

[17] C.A. Martínez-Huitle, and S. Ferro, "Electrochemical oxidation of organic pollutants for the wastewater treatment: direct and indirect processes", *Chem. Soc. Rev.,* vol. 35, no. 12, pp. 1324-1340, 2006.
[http://dx.doi.org/10.1039/B517632H] [PMID: 17225891]

[18]   Anglada, A. Urtiaga, and I. Ortiz, "Contributions of electrochemical oxidation to waste-water treatment: Fundamentals and review of applications", *J. Chem. Technol. Biotechnol.,* vol. 84, no. 12, pp. 1747-1755, 2009.
       [http://dx.doi.org/10.1002/jctb.2214]

[19]   Y. Deng, and J.D. Englehardt, "Treatment of landfill leachate by the Fenton process", *Water Res.,* vol. 40, no. 20, pp. 3683-3694, 2006.
       [http://dx.doi.org/10.1016/j.watres.2006.08.009] [PMID: 17045628]

[20]   F.L. Migliorini, N.A. Braga, S.A. Alves, M.R.V. Lanza, M.R. Baldan, and N.G. Ferreira, "Anodic oxidation of wastewater containing the Reactive Orange 16 Dye using heavily boron-doped diamond electrodes", *J. Hazard. Mater.,* vol. 192, no. 3, pp. 1683-1689, 2011.
       [http://dx.doi.org/10.1016/j.jhazmat.2011.07.007] [PMID: 21803493]

[21]   I.C. da Costa Soares, D.R. da Silva, J.H.O. do Nascimento, S. Garcia-Segura, and C.A. Martínez-Huitle, "Functional group influences on the reactive azo dye decolorization performance by electrochemical oxidation and electro-Fenton technologies", *Environ. Sci. Pollut. Res. Int.,* vol. 24, no. 31, pp. 24167-24176, 2017.
       [http://dx.doi.org/10.1007/s11356-017-0041-z] [PMID: 28884274]

[22]   V.M. Vasconcelos, C. Ponce-de-León, J.L. Nava, and M.R.V. Lanza, "Electrochemical degradation of RB-5 dye by anodic oxidation, electro-Fenton and by combining anodic oxidation–electro-Fenton in a filter-press flow cell", *J. Electroanal. Chem.,* vol. 765, pp. 179-187, 2016.
       [http://dx.doi.org/10.1016/j.jelechem.2015.07.040]

[23]   E. Yuksel, E. Gurbulak, and M. Eyvaz, "Decolorization of a reactive dye solution and treatment of a textile wastewater by electrocoagulation and chemical coagulation: Techno-economic comparison", *Environ. Prog. Sustain. Energy,* vol. 31, no. 4, pp. 524-535, 2012.
       [http://dx.doi.org/10.1002/ep.10574]

[24]   F.C. Moreira, S. Garcia-Segura, V.J.P. Vilar, R.A.R. Boaventura, and E. Brillas, "Decolorization and mineralization of Sunset Yellow FCF azo dye by anodic oxidation, electro-Fenton, UVA photoelectro-Fenton and solar photoelectro-Fenton processes", *Appl. Catal. B,* vol. 142-143, pp. 877-890, 2013.
       [http://dx.doi.org/10.1016/j.apcatb.2013.03.023]

[25]   X. Zhu, X. Chen, Z. Yang, Y. Liu, Z. Zhou, and Z. Ren, "Investigating the influences of electrode material property on degradation behavior of organic wastewaters by iron-carbon micro-electrolysis", *Chem. Eng. J.,* vol. 338, pp. 46-54, 2018.
       [http://dx.doi.org/10.1016/j.cej.2017.12.091]

[26]   A.M. Faouzi, B. Nasr, and G. Abdellatif, "Electrochemical degradation of anthraquinone dye Alizarin Red S by anodic oxidation on boron-doped diamond", *Dyes Pigments,* vol. 73, no. 1, pp. 86-89, 2007.
       [http://dx.doi.org/10.1016/j.dyepig.2005.10.013]

[27]   S. Ammar, M. Asma, N. Oturan, R. Abdelhedi, and M.A. Oturan, "Electrochemical degradation of anthraquinone dye alizarin red: Role of the electrode material", *Curr. Org. Chem.,* vol. 16, pp. 1978-1985, 2012.
       [http://dx.doi.org/10.2174/138527212803251613]

[28]   H. Hamous, A. Khenifi, Z. Bouberka, and Z. Derriche, "Electrochemical degradation of Orange G in K_2SO_4 and KCl medium", *Environ. Eng. Res.,* vol. 25, no. 4, pp. 571-578, 2020.
       [http://dx.doi.org/10.4491/eer.2019.174]

[29]   A. El-Ghenymy, F. Centellas, J.A. Garrido, R.M. Rodríguez, I. Sirés, P.L. Cabot, and E. Brillas, "Decolorization and mineralization of Orange G azo dye solutions by anodic oxidation with a boron-doped diamond anode in divided and undivided tank reactors", *Electrochim. Acta,* vol. 130, pp. 568-576, 2014.
       [http://dx.doi.org/10.1016/j.electacta.2014.03.066]

[30]   D.G. Bassyouni, H.A. Hamad, E-S.Z. El-Ashtoukhy, N.K. Amin, and M.M.A. El-Latif, "Comparative

performance of anodic oxidation and electrocoagulation as clean processes for electrocatalytic degradation of diazo dye Acid Brown 14 in aqueous medium", *J. Hazard. Mater.,* vol. 335, pp. 178-187, 2017.
[http://dx.doi.org/10.1016/j.jhazmat.2017.04.045] [PMID: 28458079]

[31]   G. Bonyadinejad, M. Sarafraz, M. Khosravi, A. Ebrahimi, S.M. Taghavi-Shahri, R. Nateghi, and S. Rastaghi, "Electrochemical degradation of the Acid Orange 10 dye on a Ti/PbO2 anode assessed by response surface methodology", *Korean J. Chem. Eng.,* vol. 33, no. 1, pp. 189-196, 2016.
[http://dx.doi.org/10.1007/s11814-015-0115-x]

[32]   M. Sarafraz, G. Khosravi, A. Bonyadinejad, and S. M. Ebrahimi, *Int. J. Environ. Health Eng.,* vol. 4, p. 31, 2015.

[33]   X. Florenza, A.M.S. Solano, F. Centellas, C.A. Martínez-Huitle, E. Brillas, and S. Garcia-Segura, "Degradation of the azo dye Acid Red 1 by anodic oxidation and indirect electrochemical processes based on Fenton's reaction chemistry. Relationship between decolorization, mineralization and products", *Electrochim. Acta,* vol. 142, pp. 276-288, 2014.
[http://dx.doi.org/10.1016/j.electacta.2014.07.117]

[34]   R. Khosravi, S. Hazrati, and M. Fazlzadeh, "Decolorization of AR18 dye solution by electrocoagulation: Sludge production and electrode loss in different current densities", *Desalination Water Treat.,* vol. 57, pp. 1-9, 2015.

[35]   D.C. Moura, M.A. Quiroz, D.R. Silva, R. Salazar, and C.A. Martínez-Huitle, "Electrochemical degradation of Acid Blue 113 dye using TiO$_2$ -nanotubes decorated with PbO$_2$ as anode", *Environ. Nanotechnol. Monit. Manag.,* vol. 5, pp. 13-20, 2016.
[http://dx.doi.org/10.1016/j.enmm.2015.11.001]

[36]   M.H. Abdel-Aziz, M. Bassyouni, M.S. Zoromba, and A.A. Alshehri, "Removal of dyes from waste solutions by anodic oxidation on an array of horizontal graphite rods anodes", *Ind. Eng. Chem. Res.,* vol. 58, no. 2, pp. 1004-1018, 2019.
[http://dx.doi.org/10.1021/acs.iecr.8b05291]

[37]   G.R. de Oliveira, N.S. Fernandes, J.V. Melo, D.R. da Silva, C. Urgeghe, and C.A. Martínez-Huitle, "Electrocatalytic properties of Ti-supported Pt for decolorizing and removing dye from synthetic textile wastewaters", *Chem. Eng. J.,* vol. 168, no. 1, pp. 208-214, 2011.
[http://dx.doi.org/10.1016/j.cej.2010.12.070]

[38]   E. Isarain-Chávez, M.D. Baró, E. Rossinyol, U. Morales-Ortiz, J. Sort, E. Brillas, and E. Pellicer, "Comparative electrochemical oxidation of methyl orange azo dye using Ti/Ir-Pb, Ti/Ir-Sn, Ti/Ru-Pb, Ti/Pt-Pd and Ti/RuO$_2$ anodes", *Electrochim. Acta,* vol. 244, pp. 199-208, 2017.
[http://dx.doi.org/10.1016/j.electacta.2017.05.101]

[39]   C. Ramírez, A. Saldaña, B. Hernández, R. Acero, R. Guerra, S. Garcia-Segura, E. Brillas, and J.M. Peralta-Hernández, "Electrochemical oxidation of methyl orange azo dye at pilot flow plant using BDD technology", *J. Ind. Eng. Chem.,* vol. 19, no. 2, pp. 571-579, 2013.
[http://dx.doi.org/10.1016/j.jiec.2012.09.010]

[40]   M. Weng, Z. Zhou, and Q. Zhang, "Electrochemical degradation of typical dyeing wastewater in aqueous solution: Performance and mechanism", *Int. J. Electrochem. Sci.,* vol. 8, no. 1, pp. 290-296, 2013.
[http://dx.doi.org/10.1016/S1452-3981(23)14020-X]

[41]   A. Mukimin, K. Wijaya, and A. Kuncaka, "Oxidation of remazolbrilliant bluer(RB.19) with *in situ* electro-generated active chlorine using Ti/PbO$_2$ electrode", *Separ. Purif. Tech.,* vol. 95, pp. 1-9, 2012.
[http://dx.doi.org/10.1016/j.seppur.2012.04.015]

[42]   C.K.C. Araújo, G.R. Oliveira, N.S. Fernandes, C.L.P.S. Zanta, S.S.L. Castro, D.R. da Silva, and C.A. Martínez-Huitle, "Electrochemical removal of synthetic textile dyes from aqueous solutions using Ti/Pt anode: Role of dye structure", *Environ. Sci. Pollut. Res. Int.,* vol. 21, no. 16, pp. 9777-9784, 2014.

[http://dx.doi.org/10.1007/s11356-014-2918-4] [PMID: 24801286]

[43]   M.A. Tabarra, H.A. Mallah, and M.M. El Jamal, "Anodic oxidation of anionic Xanthene dyes at Pt and Bdd electrodes", *J. Chem.Technol. Metal.,* vol. 49, pp. 247-253, 2014.

[44]   D.M. Araújo, C. Sáez, C.A. Martínez-Huitle, P. Cañizares, and M.A. Rodrigo, "Influence of mediated processes on the removal of Rhodamine with conductive-diamond electrochemical oxidation", *Appl. Catal. B,* vol. 166-167, pp. 454-459, 2015.
[http://dx.doi.org/10.1016/j.apcatb.2014.11.038]

[45]   A. Baddouh, G.G. Bessegato, M.M. Rguiti, B. El Ibrahimi, L. Bazzi, M. Hilali, and M.V.B. Zanoni, "Electrochemical decolorization of Rhodamine B dye: Influence of anode material, chloride concentration and current density", *J. Environ. Chem. Eng.,* vol. 6, no. 2, pp. 2041-2047, 2018.
[http://dx.doi.org/10.1016/j.jece.2018.03.007]

[46]   E. Hmani, Y. Samet, and R. Abdelhédi, "Electrochemical degradation of auramine-O dye at boron-doped diamond and lead dioxide electrodes", *Diamond Related Materials,* vol. 30, pp. 1-8, 2012.
[http://dx.doi.org/10.1016/j.diamond.2012.08.003]

[47]   A. Thiam, I. Sirés, J.A. Garrido, R.M. Rodríguez, and E. Brillas, "Decolorization and mineralization of Allura Red AC aqueous solutions by electrochemical advanced oxidation processes", *J. Hazard. Mater.,* vol. 290, pp. 34-42, 2015.
[http://dx.doi.org/10.1016/j.jhazmat.2015.02.050] [PMID: 25734532]

[48]   S. Aoudj, A. Khelifa, N. Drouiche, M. Hecini, and H. Hamitouche, "Electrocoagulation process applied to wastewater containing dyes from textile industry", *Chem. Eng. Process.,* vol. 49, no. 11, pp. 1176-1182, 2010.
[http://dx.doi.org/10.1016/j.cep.2010.08.019]

[49]   R. Li, T. Li, Y. Wan, X. Zhang, X. Liu, R. Li, H. Pu, T. Gao, X. Wang, and Q. Zhou, "Efficient decolorization of azo dye wastewater with polyaniline/graphene modified anode in microbial electrochemical systems", *J. Hazard. Mater.,* vol. 421, p. 126740, 2022.
[http://dx.doi.org/10.1016/j.jhazmat.2021.126740]

[50]   K. Govindan, Y. Oren, and M. Noel, "Effect of dye molecules and electrode material on the settling behavior of flocs in an electrocoagulation induced settling tank reactor (EISTR)", *Separ. Purif. Tech.,* vol. 133, pp. 396-406, 2014.
[http://dx.doi.org/10.1016/j.seppur.2014.04.046]

[51]   K. Govindan, M. Raja, S.U. Maheshwari, and M. Noel, "Analysis and understanding of amido black 10B dye degradation in aqueous solution by electrocoagulation with the conventional oxidants peroxomonosulfate, peroxodisulfate and hydrogen peroxide", *Environ. Sci. Water Res. Technol.,* vol. 1, no. 1, pp. 108-119, 2015.
[http://dx.doi.org/10.1039/C4EW00030G]

[52]   S.A. Carmalin, V.M. Bhalambaal, and E.C. Lima, "Re-use of carbon rods from used batteries as cathode for textile azo dye degradation in a microbial fuel cell", *Desalination Water Treat.,* vol. 79, pp. 322-328, 2017.
[http://dx.doi.org/10.5004/dwt.2017.20820]

[53]   W. Huang, J. Chen, Y. Hu, J. Chen, J. Sun, and L. Zhang, "Enhanced simultaneous decolorization of azo dye and electricity generation in microbial fuel cell (MFC) with redox mediator modified anode", *Int. J. Hydrogen Energy,* vol. 42, no. 4, pp. 2349-2359, 2017.
[http://dx.doi.org/10.1016/j.ijhydene.2016.09.216]

[54]   C. Comninellis, "Electrocatalysis in the electrochemical conversion/combustion of organic pollutants for waste water treatment", *Electrochim. Acta,* vol. 39, no. 11-12, pp. 1857-1862, 1994.
[http://dx.doi.org/10.1016/0013-4686(94)85175-1]

[55]   C.A. Martínez-Huitle, and E. Brillas, "Decontamination of wastewaters containing synthetic organic dyes by electrochemical methods: A general review", *Appl. Catal. B,* vol. 87, no. 3-4, pp. 105-145, 2009.

[http://dx.doi.org/10.1016/j.apcatb.2008.09.017]

[56]   P. Kaur, J.P. Kushwaha, and V.K. Sangal, Evaluation and disposability study of actual textile wastewater treatment by electro-oxidation method using Ti/RuO₂ anode", *Process Saf. Environ. Prot.,* vol. 111, pp. 13-22, 2017.
       [http://dx.doi.org/10.1016/j.psep.2017.06.004]

[57]   N. Bakaraki Turan, H. Sari Erkan, F. İlhan, G. Onkal Engin, and N. BakarakiTuran, "Highlighting the cathodic contribution of an electrooxidation post-treatment study on decolorization of textile wastewater effluent pre-treated with a lab-scale moving bed-membrane bioreactor", *Environ. Sci. Pollut. Res. Int.,* vol. 28, no. 20, pp. 25972-25983, 2021.
       [http://dx.doi.org/10.1007/s11356-021-12409-8] [PMID: 33479878]

[58]   N.A. Braga, C.A.A. Cairo, J.T. Matsushima, M.R. Baldan, and N.G. Ferreira, "Diamond/porous titanium three-dimensional hybrid electrodes", *J. Solid State Electrochem.,* vol. 14, no. 2, pp. 313-321, 2010.
       [http://dx.doi.org/10.1007/s10008-009-0855-9]

[59]   E. Tsantaki, T. Velegraki, A. Katsaounis, and D. Mantzavinos, "Anodic oxidation of textile dyehouse effluents on boron-doped diamond electrode", *J. Hazard. Mater.,* vol. 207-208, pp. 91-96, 2012.
       [http://dx.doi.org/10.1016/j.jhazmat.2011.03.107] [PMID: 21530081]

[60]   M. Panizza, *Importance of electrode material in the electrochemical treatment of wastewater containing organic pollutants.* Electrochem. Environ., Springer New York, 2010, pp. 25-54.
       [http://dx.doi.org/10.1007/978-0-387-68318-8_2]

[61]   E. Chatzisymeon, N.P. Xekoukoulotakis, A. Coz, N. Kalogerakis, and D. Mantzavinos, "Electrochemical treatment of textile dyes and dyehouse effluents", *J. Hazard. Mater.,* vol. 137, no. 2, pp. 998-1007, 2006.
       [http://dx.doi.org/10.1016/j.jhazmat.2006.03.032] [PMID: 16713087]

CHAPTER 6

Z-scheme: A Photocatalysis for the Remediation of Environmental Pollutants

Suresh Kumar Pandey[1] and **Dhanesh Tiwary**[1,*]

[1] *Department of Chemistry, Indian Institute of Technology (Banaras Hindu University), Varanasi-221005, India*

Abstract: Modern artificial heterostructures control redox reactions at the catalyst's active sites by effectively separating charges and transporting excitons with the help of light sources. Regarding environmental remediation, the Z-scheme—particularly in the degradation and mineralization of organic pollutants—plays a crucial role. Appropriately designed photocatalysts with Z-scheme have several benefits over conventional photocatalytic processes, including improved charge separation and effective redox process management in response to visible light. It provides the way for the creation of newer and more effective photocatalysts because it is said to make reduction and oxidation processes easier than with the constituent single precursor. In contrast to other heterostructure schemes like the Type-I and Type-II schemes, heterostructures with the Z-scheme mechanism attracted a lot of attention.

Keywords: Conduction band, Degradation, Excitons, Electron mediator, Holes, Heterojunction, Photocatalyst, Pollutants, Redox process, Semiconductor, Solar light, Valence band, Z-scheme.

INTRODUCTION

The intoxication of water poses a severe threat to life on Earth as a result of the regular growth in population and social processes. Numerous significant issues, including skin conditions, diarrhoea, dengue, malaria, and more, are brought on by contaminated water. Water is necessary for life, yet it has since turned into a deadly liquid. According to a WHO report, poor water quality is to blame for 80% of diseases in the entire world. Effluents from the food, leather, textile, and chemical industries are just a few examples of the many sources of water contamination [1 - 3]. Thus, it is essential to create efficient methods for eliminating organic material from water streams. Although there are several established classical methods for treating water, including filtration, adsorption,

* **Corresponding author Dhanesh Tiwary:** Department of Chemistry, Indian Institute of Technology (Banaras Hindu University), Varanasi-221005, India; E-mail: jagan3311d@gmail.com

Paulpandian Muthu Mareeswaran & Jegathalaprathaban Rajesh (Eds.)

and biological treatment, they all have significant shortcomings [4, 5]. One of the continually developing green technologies, photocatalysis can address both energy and environmental issues at once [6]. The photocatalysis method can also be used to reduce CO_2 and produce fuels, such as hydrogen from water splitting, in addition to degrading contaminants [7 - 17].

Semiconductor photocatalysts are responsible for all these reactions. In 1972, Fujishima used TiO_2 to create H_2 from water, which is when he first coined the phrase "photocatalysis" [18]. Since then, semiconductor photocatalysis has become the method of choice among chemists due to its special ability to directly employ solar light and the reaction's ability to take place at room temperature [19]. During photocatalysis, holes are left in the semiconductor's valence band (VB) as the electrons that were previously present are driven to their conduction band (CB) by absorbing sufficiently energetic photons. The photocatalysts must have the correct energy levels of their CB and VB as well as the desired reduction and oxidation potential to generate superoxide $(O_2^{\cdot-})$ $(EO_2/O_{2\cdot.} = -0.33$ eV) and hydroxyl $(OH^{\cdot})$ $(EOH-/OH\bullet =2.30$ eV) radicals, which are required to start the photocatalytic degradation process [20 - 22]. However, it is exceedingly difficult for a single semiconductor (type I) to have both oxidation and reduction potentials to simultaneously straddle the redox process, since, the bandgap of the semiconductor is so wide that the light response will be reduced. Additionally, individual semiconductors experience a high rate of photo-generated charge carrier recombination, which ultimately reduces the photocatalytic activity. Numerous studies have been conducted on the photocatalytic activity of certain typical semiconductor metal oxide photocatalysts, including TiO_2, ZnO, WO_3, and Ta_2O_5 [23 - 26]. However, the high recombination rate of photo-generated excitons (e^-/h^+), poor utilization of visible light, and low quantum yield further place limitations on the photocatalytic approaches using metal oxides. Scientists therefore made numerous modifications to the semiconductor (SC) in order to enhance its photocatalytic performance. The three basic components of the photocatalytic process are; (1) broad light absorption; (2) effective charge separation and their quick migration at the surface of the photocatalysts; and (3) surface catalytic redox process [27].

Despite the enormous advancements made in the science of photocatalysis over the past few decades, it still has a low photocatalytic efficiency due to the rapid recombination of photo-generated excitons which is reducing the semiconductor efficiency to utilize solar energy [28 - 31]. Therefore, for the scientists to construct this doping, loading of metal, and heterojunction, we need to develop advanced photocatalytic techniques to defeat the photoconversion process. Due to its capacity for spatial charge separation between excitons, a heterojunction type II photocatalyst is created when two semiconductors come together [29 - 32]. This catalyst exhibits efficient photocatalytic activity. The band alignment and schematic pathway of charge flow in typical type II photocatalysts are shown in Fig. (**1**). This states the migration of photo-generated electrons from the conduction band (CB) of SC-II to the CB of SC-I and simultaneously photo-generated holes travel from the valence band (VB) of SC-I to the VB of SC-II [33 - 35]. Fig. (**1**) shows that the CB of SC-I and VB of SC-II have accumulated electrons and holes, respectively.

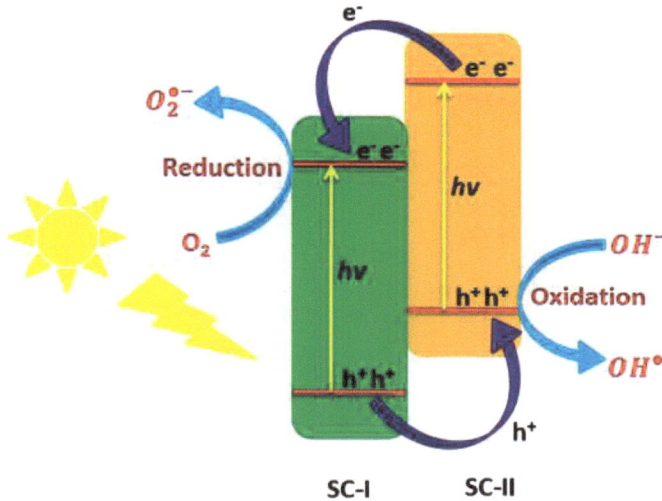

Fig. (1). A schematic representation of Type-II photocatalytic system.

As a result, effective spatial charge separation can be established to increase the photocatalytic activity of the type II photocatalytic system. Although type II systems achieve efficient charge separation, there are still some significant drawbacks, including lower redox ability and difficulty in charge migration (electrons and holes) due to electrostatic repulsion between electrons (e⁻) in the CB of SC-I and SC-II as well as holes (h⁺) in the VB of both semiconductors [36, 37]. The direct Z-scheme photocatalysis hence gained attention in 2013 when Yu *et al.* explained the remarkable photocatalytic performance of formaldehyde by g-C_3N_4/TiO_2 heterojunction photocatalysts [38, 39]. Direct Z-scheme and type II systems have been found to be fundamentally comparable, with their main difference being the direction of charge migration. The photo-generated electron in the CB of SC-I recombines with the holes existing in the VB of semiconductor SC-II in the direct Z-scheme, in contrast to type II, by the action of electrostatic attraction. Thus, direct Z-scheme photocatalysts have much better redox ability than type II photocatalytic systems because the electrons with high reduction potential and holes with strong oxidation potential remain intact in the CB of SC-

II and VB of SC-I, respectively. In addition, charge migration is also possible in case of direct Z-scheme due to the attraction between electrons and holes (electrostatic attraction). Z-scheme photocatalysts follow a charge migration pattern that closely resembles the English letter "Z" in the alphabet. As a result, they are particularly effective for the photocatalytic removal of contaminants from wastewater because they efficiently separate and migrate photo-generated electrons and holes, just like in a direct Z-scheme photocatalytic system. In this article, we examined the history and foundations of the Z-scheme photocatalytic system. We also present some tables with recent research results based on the Z-scheme system.

DEVELOPMENTS OF DIFFERENT TYPES OF Z-SCHEME PHOTOCATALYTIC SYSTEMS

Z-scheme photocatalytic systems are largely divided into two groups based on the presence of the electron mediator (Fig. **2**). Each form of Z-scheme system is covered in detail in its own section below.

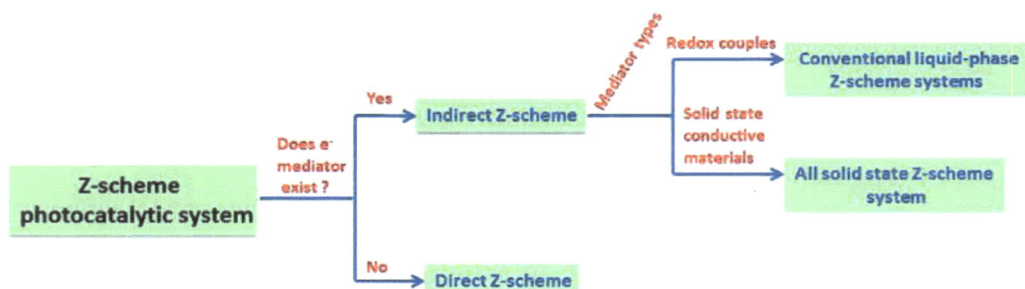

Fig. (**2**). Schematic classification of different types of Z-scheme system.

Indirect Z-scheme Photocatalytic Systems

The primary components of indirect Z-scheme photocatalysts are two different kinds of semiconductors and an electron mediator. Indirect Z-scheme photocatalysts are primarily split into two kinds based on the type of the electron mediator species involved: (1) typical liquid phase Z-scheme system: (2) A Z-scheme system that is entirely made of solid-state. These photocatalytic devices have been extensively investigated and examined in the treatment of wastewater due to the constrained rate of charge recombination and greater redox activity. Table **1** lists the ongoing research on indirect Z-scheme systems, mostly for the photocatalytic treatment of water.

Table 1. Photocatalytic performance of various photocatalysts under different kinds of illuminating sources of light.

Sr. No.	Photocatalyst	Electron Mediator	Mechanism	Pollutant	Light Source	Removal (%)	Refs.
1	g-C_3N_4/rGO/TiO_2	rGO	ASS Z-scheme	MB	Xe lamp (300 W)	92	[41]
2	g-C_3N_4@Bi/BiOBr	Bi	ASS Z-scheme	RhB	Visible light	98	[42]
3	AgI/Ag/Bi_3TaO_7	Ag	ASS Z-scheme	SMZ	Xe lamp (300 W)	98	[43]
4	CdS/AgBr/rGO	rGO	ASS Z-scheme	RhB	Xe lamp (500 W)	96	[44]
5	BiOBr-Ag-PPy	Ag	ASS Z-scheme	MG	Visible light (150 W halogen lamp)	97	[45]

(Where MB: methylene blue, RhB: rhodamine B, TC: tetracycline, SMZ: sulfamethoxazol, CIP: ciprofloxacin).

Conventional Liquid-Phase Z-Scheme Systems (1st Generation Z-scheme Photocatalysts)

In aqueous conditions, conventional Z-scheme photocatalytic systems consist of two different kinds of photocatalysts and a redox mediator. Both the VB and CB of one semiconductor should be higher than the other due to how the band locations of the semiconductors are positioned. Z-scheme relies heavily on redox mediators because both semiconductors are not near one another. A good redox mediator should have the following qualities: (1) has a favourable redox potential; (2) is transparent to light; and (3) is recyclable. The electron acceptor and donor features of the redox mediators predominate (Fig. **3**). The first-generation Z-scheme photocatalysts are effective, but they also have some major flaws. For example, because they use the redox mediators Fe^{+2}/Fe^{+3} and I^-/IO_3^- they frequently undergo reversible reactions, which reduce the photocatalyst's conversion efficiency. Additionally, the utilization of these first-generation photocatalytic systems is only applicable in the liquid phase; they are not applicable in the solid or gaseous states.

All Solid-State Z-scheme System (ASS-2nd Generation Z-scheme Photocatalysts)

Tada *et al.*, in 2006, first time introduced the concept of an all-solid-state Z-scheme (ASS Z-scheme), by developing CdS-Au-TiO_2 heterojunction photocatalyst wherein CdS and TiO_2 semiconductors are connected by Au which act as an electron mediator [40]; the structure of the ASS Z-scheme system is almost like the conventional liquid phase Z-scheme system; they differ only in terms of the type of electron mediator used. Instead of using a redox mediator like in a traditional Z-scheme, noble metals with strong photo-absorption properties or other conducting materials are employed in ASS Z-schemes as an electron bridge.

Therefore, as illustrated in Fig. (**4**), an electron mediator connects SC-I and SC-II in the ASS Z-scheme system. Rapid interfacial charge transport is facilitated by the applied electron mediator's conducting nature [38]. However, the usage of this system is primarily constrained by the high price of the noble metal and the insufficient lighting available for the catalysts.

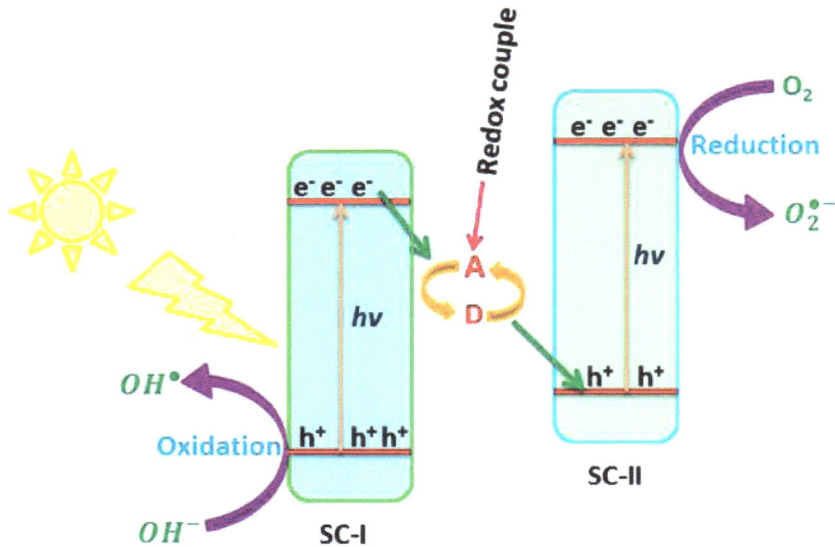

Fig. (3). A schematic diagram of Conventional Z-scheme system.

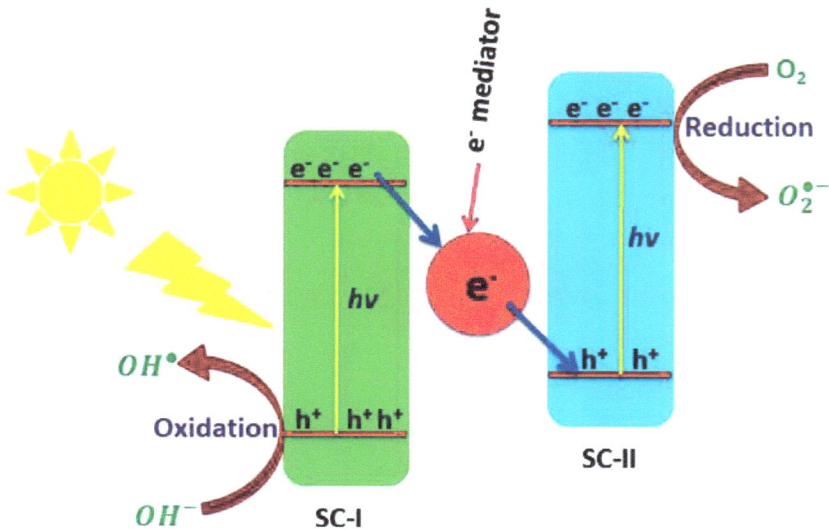

Fig. (4). ASS Z-scheme system.

Direct Z-scheme System (3rd Generation Z-scheme Photocatalysts)

The idea of a direct Z-scheme photocatalytic system (third generation) was put up in 2013 by Yu *et al*. They claim that the essential components of this kind of photocatalytic system are two diverse semiconductor types that are in close contact with one another. Direct Z-scheme photocatalysts with significantly better charge separation and photocatalytic efficiency inherit the benefits of the traditional Z-scheme and ASS Z-scheme. The cost of this photocatalytic system is significantly lowered because the direct Z-scheme photocatalyst does not require any kind of redox or electron mediator. Additionally, in the direct Z-scheme photocatalytic system, the issue of light shielding brought on by noble metals in the ASS Z-scheme is also resolved. Additionally, at the interface of their contact, direct Z-scheme devices produce an internal electric field that aids in the recombination of photo-generated electrons and holes that lack sufficient redox potential. Therefore, the excitons with higher redox potentials are not destroyed. The charge migration process of the direct Z-scheme photocatalytic system, as shown in Fig. (**5**), varies from the type II system.

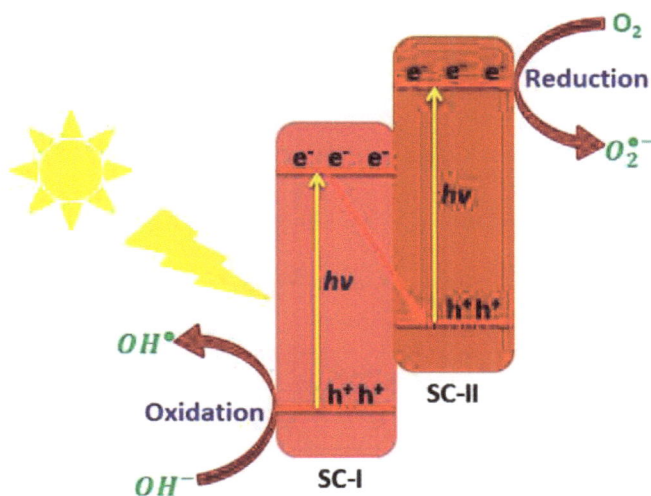

Fig. (5). Schematic representation of direct Z-scheme system.

Z-SCHEME PHOTOCATALYSTS FOR THE REMOVAL OF POLLUTANTS

Because of their capacity to segregate photo-generated charge carriers effectively and to support high redox potential, Z-scheme photocatalytic systems were discovered to be the most efficient in removing pollutants from the aqueous system. Additionally, under the illumination of solar lights, it particularly generates holes, superoxide, and hydroxyl radicals during the photocatalytic activities. By using oxidative or reductive pathways, these photo-generated species act as reaction intermediates and break down the contaminants.

Oxidative Removal of the Pollutant

X. Yue *et al.* [46] successfully synthesized the photocatalysts m-Bi$_2$O$_4$/NCDs. It was discovered that the hydrothermal approach destroys both pollutants in 30 and 120 minutes, respectively, when its photocatalytic activity was tested on MO and phenol. When the activity of m-Bi$_2$O$_4$/NCDs and pure m-Bi$_2$O$_4$ were evaluated individually, the authors reported that the hotocatalytic degradation efficiency of m-Bi$_2$O$_4$/NCDs is higher (0.12458 min^{-1}) than the pure m-Bi$_2$O$_4$ (0.05368 min^{-1}). To explain the increased photocatalytic efficiency attained by m-Bi$_2$O$_4$/NCDs, the authors postulated a mechanism based on the Z-scheme pathway. According to the hypothesized mechanism, m-Bi$_2$O$_4$ and NCDs come together to create a heterojunction. The CB and VB of m-Bi$_2$O$_4$ and NCDs, contain photogenerated electrons and holes, respectively. The effective charge (e$^-$/h$^+$) separation in this fashion is due to the electrons in the CB of NCDs and the holes in the VB of m-Bi$_2$O$_4$ remain intact even in the presence of opposite charge attraction. Furthermore, the increased photocatalytic activity of the m-Bi$_2$O$_4$/NCDs catalysts is due to holes in the VB of m-Bi$_2$O$_4$ and superoxide radical produced by the reduction of oxygen. The detailed evaluation of the photocatalytic activity achieved by various photocatalysts through direct Z-scheme pathways is given in Tables **1** and **2**.

Table 2. Photocatalytic performance of various photocatalysts under different kinds of illuminating sources of light.

Sr. No.	Photocatalysts	Mechanism	Pollutant	Radical Type	Experimental Condition	Removal (%)	Refs.
1	Bi$_2$O$_3$/SnO$_2$	Direct Z-scheme	BPA	O$_2$$^{\bullet-}$, h$^-$	Visible light (350 WXe lamp)	93	[47]
2	CdTe/TiO$_2$	Direct Z-scheme	TC	O$_2$$^{\bullet-}$, OH$^{\bullet}$, h$^-$	Visible light (400 W halogen lamp)	78	[48]
3	CuS/WO$_3$	Direct Z-scheme	RhB	h$^+$, O$_2$$^{\bullet-}$	Visible light (500 W Xe lamp)	95	[49]
4	g-C$_3$N$_4$/BiMoO$_6$	Direct Z-scheme	MB	O$_2$$^{\bullet-}$, OH$^{\bullet}$, h$^+$	Visible light (50 W LED)	90	[50]
5	TiO$_2$-x/AgI	Direct Z-scheme	RhB	O$_2$$^{\bullet-}$, OH$^{\bullet}$	Visible light (300 WXe lamp)	96	[51]
6	V$_2$O$_5$/g-C$_3$N$_4$	Direct Z-scheme	RhB	O$_2$$^{\bullet-}$, h$^-$	Visible light (250 WXe lamp)	95	[52]
7	Bi$_2$WO$_6$/CuBi$_2$O$_4$	Direct Z-scheme	TC	O$_2$$^{\bullet-}$, h$^-$	Visible light (300 W Xe lamp)	94	[53]

(Table 2) cont.....

Sr. No.	Photocatalysts	Mechanism	Pollutant	Radical Type	Experimental Condition	Removal (%)	Refs.
8	BiOI-Bi$_2$O$_4$	Direct Z-scheme	RhB	O$_2$$^{\bullet-}$, h$^+$	Visible light (100 W LED)	97	[54]
9	AgI/Bi$_4$V$_2$O$_{11}$	Direct Z-scheme	Sulfamethazine	O$_2$$^{\bullet-}$, h$^+$	Visible light	93	[55]

(BPA: Bisphenol A, TC: Tetracycline, RhB: Rhodamine B, MB: Methylene blue, MG: Malachite green).

CONCLUDING REMARKS

Based on the research works discussed above, we can infer that Z-scheme photocatalysts are particularly effective in photocatalytic processes because they can harvest solar light and separate the relevant photo-generated electron and holes with strong redox potential. The ASS Z-scheme and direct Z-scheme photocatalytic systems are extensively researched and examined in many environmental issues and energy storage applications, among the numerous types of Z-scheme systems covered above. In its mature form, Z-scheme photocatalysts have the potential to be a catalyst for the removal of environmental contaminants. Additionally, it was discovered that the majority of composite photocatalysts are effective in removing several pollutants.

ACKNOWLEDGEMENT

The authors acknowledge IIT (BHU), Varanasi for providing an internet facility.

REFERENCES

[1]   S. Sarmah, and A. Kumar, "Photocatalytic activity of polyaniline-TiO$_2$ nanocomposites", *Indian J. Phys.,* vol. 85, no. 5, pp. 713-726, 2011.
[http://dx.doi.org/10.1007/s12648-011-0071-1]

[2]   G. Crini, H. Peindy, F. Gimbert, and C. Robert, "Removal of C.I. Basic Green 4 (Malachite Green) from aqueous solutions by adsorption using cyclodextrin-based adsorbent: Kinetic and equilibrium studies", *Separ. Purif. Tech.,* vol. 53, no. 1, pp. 97-110, 2007.
[http://dx.doi.org/10.1016/j.seppur.2006.06.018]

[3]   W. Cheng, S-G. Wang, L. Lu, W-X. Gong, X-W. Liu, B-Y. Gao, and H-Y. Zhang, "Removal of malachite green (MG) from aqueous solutions by native and heat-treated anaerobic granular sludge", *Biochem. Eng. J.,* vol. 39, no. 3, pp. 538-546, 2008.
[http://dx.doi.org/10.1016/j.bej.2007.10.016]

[4]   A. Mittal, "Adsorption kinetics of removal of a toxic dye, Malachite Green, from wastewater by using hen feathers", *J. Hazard. Mater.,* vol. 133, no. 1-3, pp. 196-202, 2006.
[http://dx.doi.org/10.1016/j.jhazmat.2005.10.017] [PMID: 16326001]

[5]   J. Wang, M. Qiao, K. Wei, J. Ding, Z. Liu, K.Q. Zhang, and X. Huang, "Decolorizing activity of malachite green and its mechanisms involved in dye biodegradation by Achromobacter xylosoxidans MG1", *Microbial Physiology,* vol. 20, no. 4, pp. 220-227, 2011.
[http://dx.doi.org/10.1159/000330669] [PMID: 21865764]

[6]   G. Palmisano, V. Augugliaro, M. Pagliaro, and L. Palmisano, "Photocatalysis: A promising route for

21st century organic chemistry", *Chem. Commun.,* no. 33, pp. 3425-3437, 2007.
[http://dx.doi.org/10.1039/b700395c] [PMID: 17700873]

[7]   C.H. Liao, C.W. Huang, and J.C.S. Wu, "Hydrogen production from semiconductor-based photocatalysis *via* water splitting", *Catalysts,* vol. 2, no. 4, pp. 490-516, 2012.
[http://dx.doi.org/10.3390/catal2040490]

[8]   T. Hisatomi, J. Kubota, and K. Domen, "Recent advances in semiconductors for photocatalytic and photoelectrochemical water splitting", *Chem. Soc. Rev.,* vol. 43, no. 22, pp. 7520-7535, 2014.
[http://dx.doi.org/10.1039/C3CS60378D] [PMID: 24413305]

[9]   K. Maeda, and K. Domen, "Photocatalytic water splitting: Recent progress and future challenges", *J. Phys. Chem. Lett.,* vol. 1, no. 18, pp. 2655-2661, 2010.
[http://dx.doi.org/10.1021/jz1007966]

[10]   M.R.D. Khaki, M.S. Shafeeyan, A.A.A. Raman, and W.M.A.W. Daud, "Application of doped photocatalysts for organic pollutant degradation - A review", *J. Environ. Manage.,* vol. 198, no. Pt 2, pp. 78-94, 2017.
[http://dx.doi.org/10.1016/j.jenvman.2017.04.099] [PMID: 28501610]

[11]   D. Zhu, and Q. Zhou, "Action and mechanism of semiconductor photocatalysis on degradation of organic pollutants in water treatment: A review", *Environ. Nanotechnol. Monit. Manag.,* vol. 12, p. 100255, 2019.
[http://dx.doi.org/10.1016/j.enmm.2019.100255]

[12]   W. Zou, B. Gao, Y.S. Ok, and L. Dong, "Integrated adsorption and photocatalytic degradation of volatile organic compounds (VOCs) using carbon-based nanocomposites: A critical review", *Chemosphere,* vol. 218, pp. 845-859, 2019.
[http://dx.doi.org/10.1016/j.chemosphere.2018.11.175] [PMID: 30508803]

[13]   Z. Shayegan, C.S. Lee, and F. Haghighat, "TiO$_2$ photocatalyst for removal of volatile organic compounds in gas phase – A review", *Chem. Eng. J.,* vol. 334, pp. 2408-2439, 2018.
[http://dx.doi.org/10.1016/j.cej.2017.09.153]

[14]   S. Nahar, M. Zain, A. Kadhum, H. Hasan, and M. Hasan, "Advances in photocatalytic CO$_2$ reduction with water: A review", *Materials,* vol. 10, no. 6, p. 629, 2017.
[http://dx.doi.org/10.3390/ma10060629] [PMID: 28772988]

[15]   Z. Xiong, Z. Lei, Y. Li, L. Dong, Y. Zhao, and J. Zhang, "A review on modification of facet-engineered TiO$_2$ for photocatalytic CO$_2$ reduction", *J. Photochem. Photobiol. Photochem. Rev.,* vol. 36, pp. 24-47, 2018.
[http://dx.doi.org/10.1016/j.jphotochemrev.2018.07.002]

[16]   W. Wang, G. Huang, J.C. Yu, and P.K. Wong, "Advances in photocatalytic disinfection of bacteria: Development of photocatalysts and mechanisms", *J. Environ. Sci.,* vol. 34, pp. 232-247, 2015.
[http://dx.doi.org/10.1016/j.jes.2015.05.003] [PMID: 26257366]

[17]   C. Zhang, Y. Li, D. Shuai, Y. Shen, and D. Wang, "Progress and challenges in photocatalytic disinfection of waterborne Viruses: A review to fill current knowledge gaps", *Chem. Eng. J.,* vol. 355, pp. 399-415, 2019.
[http://dx.doi.org/10.1016/j.cej.2018.08.158]

[18]   A. Fujishima, and K. Honda, "Electrochemical photolysis of water at a semiconductor electrode", *Nature,* vol. 238, no. 5358, pp. 37-38, 1972.
[http://dx.doi.org/10.1038/238037a0] [PMID: 12635268]

[19]   S.M. Gupta, and M. Tripathi, "An overview of commonly used semiconductor nanoparticles in photocatalysis", *High Energy Chem.,* vol. 46, no. 1, pp. 1-9, 2012.
[http://dx.doi.org/10.1134/S0018143912010134]

[20]   A.H. Mamaghani, F. Haghighat, and C.S. Lee, "Photocatalytic oxidation technology for indoor environment air purification: The state-of-the-art", *Appl. Catal. B,* vol. 203, pp. 247-269, 2017.

[http://dx.doi.org/10.1016/j.apcatb.2016.10.037]

[21] C. Zhang, T. Li, J. Zhang, S. Yan, and C. Qin, "Degradation of p-nitrophenol using a ferrous-tripolyphosphate complex in the presence of oxygen: The key role of superoxide radicals", *Appl. Catal. B,* vol. 259, p. 118030, 2019.
[http://dx.doi.org/10.1016/j.apcatb.2019.118030]

[22] Y. Fu, T. Huang, L. Zhang, J. Zhu, and X. Wang, "Ag/g-C 3 N 4 catalyst with superior catalytic performance for the degradation of dyes: A borohydride-generated superoxide radical approach", *Nanoscale,* vol. 7, no. 32, pp. 13723-13733, 2015.
[http://dx.doi.org/10.1039/C5NR03260A] [PMID: 26220662]

[23] J. Yu, H. Yu, B. Cheng, M. Zhou, and X. Zhao, "Enhanced photocatalytic activity of TiO_2 powder (P25) by hydrothermal treatment", *J. Mol. Catal. Chem.,* vol. 253, no. 1-2, pp. 112-118, 2006.
[http://dx.doi.org/10.1016/j.molcata.2006.03.021]

[24] J. Mu, C. Shao, Z. Guo, Z. Zhang, M. Zhang, P. Zhang, B. Chen, and Y. Liu, "High photocatalytic activity of ZnO-carbon nanofiber heteroarchitectures", *ACS Appl. Mater. Interfaces,* vol. 3, no. 2, pp. 590-596, 2011.
[http://dx.doi.org/10.1021/am101171a] [PMID: 21291208]

[25] F. Amano, E. Ishinaga, and A. Yamakata, "Effect of particle size on the photocatalytic activity of WO 3 particles for water oxidation", *J. Phys. Chem. C,* vol. 117, no. 44, pp. 22584-22590, 2013.
[http://dx.doi.org/10.1021/jp408446u]

[26] X. Shi, D. Ma, Y. Ma, and A. Hu, "N-doping Ta_2O_5 nanoflowers with strong adsorption and visible light photocatalytic activity for efficient removal of methylene blue", *J. Photochem. Photobiol. Chem.,* vol. 332, pp. 487-496, 2017.
[http://dx.doi.org/10.1016/j.jphotochem.2016.09.014]

[27] Y. Tachibana, L. Vayssieres, and J.R. Durrant, "Artificial photosynthesis for solar water-splitting", *Nat. Photonics,* vol. 6, no. 8, pp. 511-518, 2012.
[http://dx.doi.org/10.1038/nphoton.2012.175]

[28] H. Wang, L. Zhang, Z. Chen, J. Hu, S. Li, Z. Wang, J. Liu, and X. Wang, "Semiconductor heterojunction photocatalysts: Design, construction, and photocatalytic performances", *Chem. Soc. Rev.,* vol. 43, no. 15, pp. 5234-5244, 2014.
[http://dx.doi.org/10.1039/C4CS00126E] [PMID: 24841176]

[29] S. Cao, J. Low, J. Yu, and M. Jaroniec, "Polymeric photocatalysts based on graphitic carbon nitride", *Adv. Mater.,* vol. 27, no. 13, pp. 2150-2176, 2015.
[http://dx.doi.org/10.1002/adma.201500033] [PMID: 25704586]

[30] W. Chen, T-Y. Liu, T. Huang, X-H. Liu, G-R. Duan, X-J. Yang, and S-M. Chen, A novel yet simple strategy to fabricate visible light responsive C,N-TiO_2 /g-C_3N_4 heterostructures with significantly enhanced photocatalytic hydrogen generation", *RSC Advances,* vol. 5, no. 122, pp. 101214-101220, 2015.
[http://dx.doi.org/10.1039/C5RA18302B]

[31] J. Low, B. Cheng, and J. Yu, "Surface modification and enhanced photocatalytic CO_2 reduction performance of TiO_2: A review", *Appl. Surf. Sci.,* vol. 392, pp. 658-686, 2017.
[http://dx.doi.org/10.1016/j.apsusc.2016.09.093]

[32] S.J.A. Moniz, S.A. Shevlin, D.J. Martin, Z.X. Guo, and J. Tang, "Visible-light driven heterojunction photocatalysts for water splitting – a critical review", *Energy Environ. Sci.,* vol. 8, no. 3, pp. 731-759, 2015.
[http://dx.doi.org/10.1039/C4EE03271C]

[33] H. Li, Y. Zhou, W. Tu, J. Ye, and Z. Zou, "State-of-the-Art progress in diverse heterostructured photocatalysts toward promoting photocatalytic performance", *Adv. Funct. Mater.,* vol. 25, no. 7, pp. 998-1013, 2015.
[http://dx.doi.org/10.1002/adfm.201401636]

[34]   J.L. White, M.F. Baruch, J.E. Pander III, Y. Hu, I.C. Fortmeyer, J.E. Park, T. Zhang, K. Liao, J. Gu, Y. Yan, T.W. Shaw, E. Abelev, and A.B. Bocarsly, "Light-Driven heterogeneous reduction of carbon dioxide: Photocatalysts and photoelectrodes", *Chem. Rev.,* vol. 115, no. 23, pp. 12888-12935, 2015.
[http://dx.doi.org/10.1021/acs.chemrev.5b00370] [PMID: 26444652]

[35]   M. Reza Gholipour, C.T. Dinh, F. Béland, and T.O. Do, "Nanocomposite heterojunctions as sunlight-driven photocatalysts for hydrogen production from water splitting", *Nanoscale,* vol. 7, no. 18, pp. 8187-8208, 2015.
[http://dx.doi.org/10.1039/C4NR07224C] [PMID: 25804291]

[36]   J. Xiao, Y. Xie, and H. Cao, "Organic pollutants removal in wastewater by heterogeneous photocatalytic ozonation", *Chemosphere,* vol. 121, pp. 1-17, 2015.
[http://dx.doi.org/10.1016/j.chemosphere.2014.10.072] [PMID: 25479808]

[37]   S. Sun, "Recent advances in hybrid Cu_2O-based heterogeneous nanostructures", *Nanoscale,* vol. 7, no. 25, pp. 10850-10882, 2015.
[http://dx.doi.org/10.1039/C5NR02178B] [PMID: 26059894]

[38]   P. Zhou, J. Yu, and M. Jaroniec, "All-solid-state Z-scheme photocatalytic systems", *Adv. Mater.,* vol. 26, no. 29, pp. 4920-4935, 2014.
[http://dx.doi.org/10.1002/adma.201400288] [PMID: 24888530]

[39]   Y.W. Su, W.H. Lin, Y.J. Hsu, and K.H. Wei, "Conjugated polymer/nanocrystal nanocomposites for renewable energy applications in photovoltaics and photocatalysis", *Small,* vol. 10, no. 22, pp. 4427-4442, 2014.
[http://dx.doi.org/10.1002/smll.201401508] [PMID: 25074641]

[40]   H. Tada, T. Mitsui, T. Kiyonaga, T. Akita, and K. Tanaka, "All-solid-state Z-scheme in CdS–Au–TiO_2 three-component nanojunction system", *Nat. Mater.,* vol. 5, no. 10, pp. 782-786, 2006.
[http://dx.doi.org/10.1038/nmat1734] [PMID: 16964238]

[41]   F. Wu, X. Li, W. Liu, and S. Zhang, "Highly enhanced photocatalytic degradation of methylene blue over the indirect all-solid-state Z-scheme g-C_3N_4-RGO-TiO_2 nanoheterojunctions", *Appl. Surf. Sci.,* vol. 405, pp. 60-70, 2017.
[http://dx.doi.org/10.1016/j.apsusc.2017.01.285]

[42]   H. Liu, H. Zhou, X. Liu, H. Li, C. Ren, X. Li, W. Li, Z. Lian, and M. Zhang, "Engineering design of hierarchical g-C3N4@Bi/BiOBr ternary heterojunction with Z-scheme system for efficient visible-light photocatalytic performance", *J. Alloys Compd.,* vol. 798, pp. 741-749, 2019.
[http://dx.doi.org/10.1016/j.jallcom.2019.05.303]

[43]   M. Ren, J. Chen, P. Wang, J. Hou, J. Qian, C. Wang, and Y. Ao, "Construction of silver iodide/silver/bismuth tantalate Z-scheme photocatalyst for effective visible light degradation of organic pollutants", *J. Colloid Interface Sci.,* vol. 532, pp. 190-200, 2018.
[http://dx.doi.org/10.1016/j.jcis.2018.07.141] [PMID: 30081264]

[44]   J. Zhang, Z. Zhang, W. Zhu, and X. Meng, "Boosted photocatalytic degradation of Rhodamine B pollutants with Z-scheme CdS/AgBr-rGO nanocomposite", *Appl. Surf. Sci.,* vol. 502, p. 144275, 2020.
[http://dx.doi.org/10.1016/j.apsusc.2019.144275]

[45]   X. Liu, and L. Cai, "Novel indirect Z-scheme photocatalyst of Ag nanoparticles and polymer polypyrrole co-modified BiOBr for photocatalytic decomposition of organic pollutants", *Appl. Surf. Sci.,* vol. 445, pp. 242-254, 2018.
[http://dx.doi.org/10.1016/j.apsusc.2018.03.178]

[46]   X. Yue, X. Miao, Z. Ji, X. Shen, H. Zhou, L. Kong, G. Zhu, X. Li, and S. Ali Shah, "Nitrogen-doped carbon dots modified dibismuth tetraoxide microrods: A direct Z-scheme photocatalyst with excellent visible-light photocatalytic performance", *J. Colloid Interface Sci.,* vol. 531, pp. 473-482, 2018.
[http://dx.doi.org/10.1016/j.jcis.2018.07.086] [PMID: 30055442]

[47]   L. Chu, J. Zhang, Z. Wu, C. Wang, Y. Sun, S. Dong, and J. Sun, "Solar-driven photocatalytic removal

of organic pollutants over direct Z-scheme coral-branch shape Bi_2O_3/SnO_2 composites", *Mater. Charact.*, vol. 159, p. 110036, 2020.
[http://dx.doi.org/10.1016/j.matchar.2019.110036]

[48]   Y. Gong, Y. Wu, Y. Xu, L. Li, C. Li, X. Liu, and L. Niu, All-solid-state Z-scheme $CdTe/TiO_2$ heterostructure photocatalysts with enhanced visible-light photocatalytic degradation of antibiotic waste water", *Chem. Eng. J.*, vol. 350, pp. 257-267, 2018.
[http://dx.doi.org/10.1016/j.cej.2018.05.186]

[49]   C. Song, X. Wang, J. Zhang, X. Chen, and C. Li, "Enhanced performance of direct Z-scheme CuS-WO_3 system towards photocatalytic decomposition of organic pollutants under visible light", *Appl. Surf. Sci.*, vol. 425, pp. 788-795, 2017.
[http://dx.doi.org/10.1016/j.apsusc.2017.07.082]

[50]   J. Lv, K. Dai, J. Zhang, L. Geng, C. Liang, Q. Liu, G. Zhu, and C. Chen, "Facile synthesis of Z-scheme graphitic-C_3N_4/Bi_2MoO_6 nanocomposite for enhanced visible photocatalytic properties", *Appl. Surf. Sci.*, vol. 358, pp. 377-384, 2015.
[http://dx.doi.org/10.1016/j.apsusc.2015.06.183]

[51]   J. Li, B. Liu, X. Han, B. Liu, J. Jiang, S. Liu, J. Zhang, and H. Shi, "Direct Z-scheme TiO_{2-x}/AgI heterojunctions for highly efficient photocatalytic degradation of organic contaminants and inactivation of pathogens", *Separ. Purif. Tech.*, vol. 261, p. 118306, 2021.
[http://dx.doi.org/10.1016/j.seppur.2021.118306]

[52]   Y. Hong, Y. Jiang, C. Li, W. Fan, X. Yan, M. Yan, and W. Shi, "*In-situ* synthesis of direct solid-state Z-scheme $V_2O_5/g-C_3N_4$ heterojunctions with enhanced visible light efficiency in photocatalytic degradation of pollutants", *Appl. Catal. B*, vol. 180, pp. 663-673, 2016.
[http://dx.doi.org/10.1016/j.apcatb.2015.06.057]

[53]   X. Yuan, D. Shen, Q. Zhang, H. Zou, Z. Liu, and F. Peng, "Z-scheme $Bi_2WO_6/CuBi_2O_4$ heterojunction mediated by interfacial electric field for efficient visible-light photocatalytic degradation of tetracycline", *Chem. Eng. J.*, vol. 369, pp. 292-301, 2019.
[http://dx.doi.org/10.1016/j.cej.2019.03.082]

[54]   H. Qin, K. Wang, L. Jiang, J. Li, X. Wu, and G. Zhang, "Ultrasonic-assisted fabrication of a direct Z-scheme $BiOI/Bi_2O_4$ heterojunction with superior visible light-responsive photocatalytic performance", *J. Alloys Compd.*, vol. 821, p. 153417, 2020.
[http://dx.doi.org/10.1016/j.jallcom.2019.153417]

[55]   X.J. Wen, Qian-Lu, X.X. Lv, J. Sun, J. Guo, Z.H. Fei, and C.G. Niu, "Photocatalytic degradation of sulfamethazine using a direct Z-Scheme $AgI/Bi_4V_2O_{11}$ photocatalyst: Mineralization activity, degradation pathways and promoted charge separation mechanism", *J. Hazard. Mater.*, vol. 385, p. 121508, 2020.
[http://dx.doi.org/10.1016/j.jhazmat.2019.121508] [PMID: 31732335]

CHAPTER 7

A Review of Various Materials under Different Conditions for Efficient Photocatalytic Dye Degradation

SP. Keerthana[1], R. Yuvakkumar[1,*] and G. Ravi[1]

[1] *Department of Physics, Alagappa University, Karaikudi, Tamil Nadu, India*

Abstract: Large amounts of more toxic dye water have been released into the environment recently as a result of the expansion of the textile industry. There are numerous approaches that have been found and applied to lessen the water's toxicity. One of the processes that operate when there is light illumination is photocatalysis. The electrons in the valence band absorb light illumination when exposed to it, excite the conduction band, and create a hole in the valence band. The dye compounds will be lessened by the recombination of these created electron-hole pairs. Materials for effective photocatalysis are being researched. Many factors affect the photocatalytic performance, including narrow bandgap, high surface area, and good recombination rate. TiO_2 is a semiconducting material, however, due to its higher bandgap values, it has a lower potential when exposed to light. This article provides a brief overview of several materials that can be affected by a variety of factors, such as doping, surfactant addition, and composites made of carbon-based materials. It also compares how well each material performs in terms of lowering hazardous pollutants and provides an illustration of the mechanism.

Keywords: Bandgap, Composite, Doping, Methylene blue, Malachite green, Photocatalysts, Reusage, Rhodamine B, Surfactant, TiO_2, Wastewater management.

INTRODUCTION

All living things on Earth depend primarily on water. It is the best place to go for daily requirements. 75% of the surface of the planet is covered by water sources, but just 2.5% of those are freshwater sources. 2.5% of the water source is insufficient for the entire population in our world which is expanding at an incredibly rapid rate. For humans and all other living things in the environment, it would be disastrous if 2.5% of the water were polluted [1]. The water has been

* **Corresponding author R. Yuvakkumar:** Department of Physics, Alagappa University, Karaikudi, Tamil Nadu, India; E-mail: yuvakkumarr@alagappauniversity.ac.in

Paulpandian Muthu Mareeswaran & Jegathalaprathaban Rajesh (Eds.)

contaminated by both humans and businesses. The increased expansion of industrialisation has been a major factor in the release of highly hazardous, contaminated wastewater. On the other hand, companies utilize a lot of water, and without any pretreatment, they completely release that water into the neighbouring water resources. When a material causes a change in the behaviour of water that is detrimental, this change is referred to as water pollution [2].

The wastewater in this now-industrialized area is primarily composed of organic salts, carcinogens, poisons, and untreated dyes. The primary sources of pollution today are the food, pharmaceutical, leather and tannery, and textile industries. One of the main contributors to the societal exchange of highly polluted wastewater is the textile industry. To control the water shortage, this needs to be eliminated quickly [3]. All varieties of organic, non-organic, and processed pharmaceutical and dye chemicals can be found in wastewater as pollutants. The main elements that have been found in the wastewater are textile dyes. There are 100,000 different dyes, and 7×10^5 tonnes of dyes produced each year. Dyes are stable materials that will form a solid bond with the host material, and they are extremely difficult to remove [4]. Due to their complex structural makeup, the dye strains are challenging to remove. The benefit of dye products is their rich colorization. The substances that have auxochromes and chromophores, which are responsible for colour and colour intensity, are called dyes. Auxochrome will determine how intense the light is, and the chromophore will determine how colourful it is. The dyes were divided into natural, synthetic, azo, diazo, anthraquinone, direct, dispersion, vat, sulphur, solvent, reactive, *etc.* based on the source, chromophore type, and substrate [5]. The major issue right now is that these organic materials are thrown into water resources and the ground, which is highly bad for the ecology. These dangerous substances will undoubtedly have negative consequences on both aquatic life and people. The Biological Oxygen Demand (BOD) and Chemical Oxygen Demand (COD) will change as a result of the organic substances being deposited into water sources. Aquatic animals will suffer from negative impacts from them, which will cause their demise [6]. The dissolved solids, BOD, COD, colours, and hazardous substances make up the textile wastewater. The percentages of BOD, COD, pH, TOC, and temperature are included in the standard value for water quality checks. There are no acceptable water quality guidelines from before the 20th century. The parameters are introduced by the rise in effluents. Researchers have developed a variety of physical, chemical, and biological strategies to address these issues. In the past, physical processes like reverse osmosis, ion exchange, nanofiltration, coagulation, adsorption, and radiation were used. Following biological techniques like enzyme degradation and microbial adsorption. It was pursued to use chemical processes such as oxidation, ozonation, improved oxidation processes, electrochemical destruction, and the Fenton reaction, as well as ultraviolet light. Physical measures were no longer used because of their lengthy processes and ineffectiveness in eliminating novel dyes with aggressive behaviours. The best technique for efficiently degrading harmful contaminants was the Advanced Oxidation Process (AOP) [7]. Here, we discuss several AOPs and materials with a range of characteristics to effectively remove pollutants.

ADVANCED OXIDATION PROCESSES (AOPS)

AOPs are effective chemical processes that completely purify the water for reuse. Complete compound mineralization will result from this method, which is suited for both harmful contaminants and chemically stable substances. Fig. (1) displays the various AOP processes. AOPs can be classified as homogeneous or heterogeneous techniques. In addition, there are two categories for homogenous methods: with and without energy. The processes O_3/UV, H_2O_2/UV, $O_3/H_2O_2/UV$, electrochemical oxidation, electro-Fenton, and photo-Fenton all required energy. Heterogeneous photocatalysis, photocatalytic ozonation, and catalytic ozonation make up the heterogeneous process. Photocatalysis has the potential to be one of these processes that reduce contaminants from wastewater the most effectively [8].

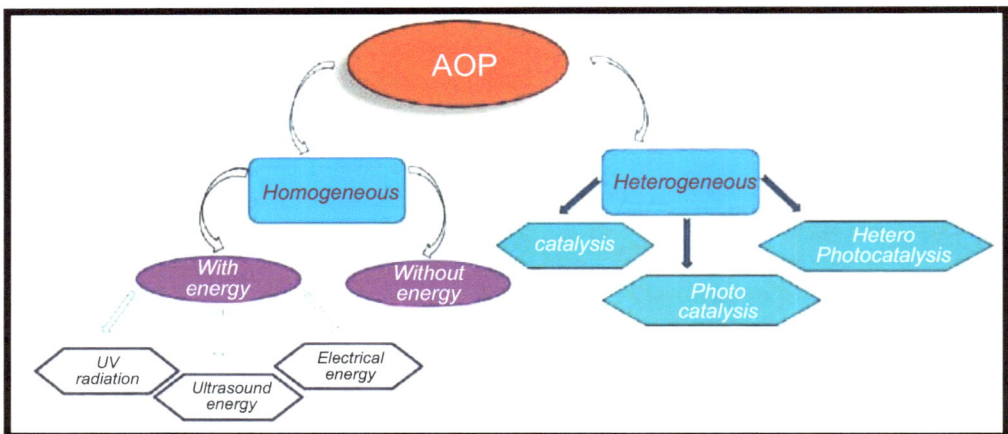

Fig. (1). Shows different types of AOP process.

When a semiconducting photocatalyst is made, photocatalysis is the process of producing electron-hole pairs while the catalyst is exposed to light. The mechanism of photocatalysis is depicted in Fig. (2). The term "photocatalyst" is also used to describe the semiconductors involved in photocatalysis. The phrase "photocatalyst" comes from the words "photon" and "catalyst," and it means that the substance will alter the rate of reaction in relation to light.

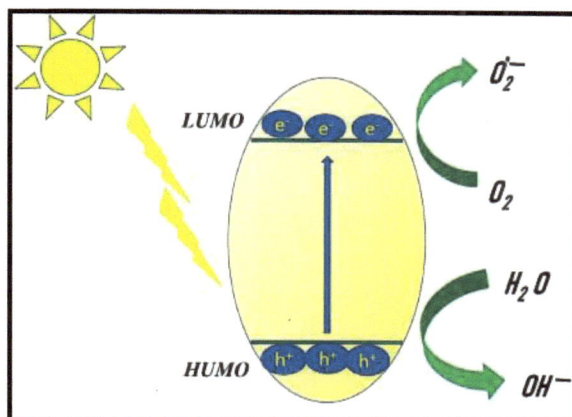

Fig. (2). Mechanism of photocatalysis.

There are two categories of photocatalysis:

i. Homogeneous photocatalysis (semiconductor and reactant in the same phase).
ii. Heterogeneous photocatalysis (Semiconductor and reactant are at different phases).

The bandgap of the materials must be understood in order to create effective photocatalysts. The difference between the material's HOMO and LUMO levels is used to compute the bandgap. Bandgaps in semiconductors are larger than 1.5 eV to 3.0 eV, in insulators, bandgaps are greater than 5.0 eV, and in metals, they are less than 1.0 eV. The bandgap of semiconductors is suitable for photocatalytic applications. When the photocatalyst is exposed to light irradiation, the electrons in the valence band absorb energy and travel to the conduction band. At the same time, holes are created in the valence band, forming electron-hole pairs that can undergo recombination to produce oxidation and reduction. The choice of light affects the photocatalyst's effectiveness. To create a respectable photocatalyst, the correct wavelength must be employed [9]. Photocatalysts were employed in a wide range of industries, including energy storage, antifouling, water management, air purification, and more [10].

MATERIALS FOR PHOTOCATALYSIS

The majority of the materials are being researched for effective photocatalysis and for their ease of water separation for further uses. For a material to be used in visible light, it must have a narrow bandgap. Utilizing UV light is not thought to be the ideal method because sunlight contains a large amount of visible light, which will make the process more affordable. The greatest choice for using visible light to reach the core end will be the narrow bandgap. To obtain the low bandgap required for photocatalysts to function under visible light, a wide range of materials has been explored here [11].

Metal Oxides-based Photocatalysts

TiO_2, CuO_2, and ZnO photocatalysts are effective in the process of degrading dyes. TiO_2 was previously employed in the process of treating wastewater. Transition metal oxides including Ni, Cu, and Zn were known to be used in photocatalytic applications and in all other commercial applications after thorough investigation. However, pure metal oxides have a wide bandgap, which has turned into the biggest drawback of the material when used in visible light [12]. Doping is one method being researched to alter the optical, chemical, and electrical properties of the metal oxides in order to eliminate this problem. When the metal oxides are modified, the efficiency is higher than with pure metal oxides [13].

Metal Sulphide-based Photocatalysts

Earth is plentiful in transition metal sulphides, and the combination of metals with sulphides improves semiconducting characteristics. Pyrites, simple-fluorite, anti-fluorite, and metal sulphides with tetragonal crystal structures are a few examples of the several types of transition metal sulphides. Better optical performance, electrical characteristics, and greater mechanical and thermal stability can all be found in transition metal sulphides [14]. It has been documented in earlier literature [15] that metal sulphides like ZnS and CdS were employed for photocatalytic applications.

Metal Ferrites-based Photocatalysts

The substances that must be mixed with iron oxide along with additional substances are known as ferrites. The formula for spinel ferrites is MFe_2O_4, where M stands for a cation and Fe_2O_4 is a substance that is magnetically stable. Ferrite materials have a tiny bandgap, making them more effective for photocatalysis using solar light [16]. The materials that have been extensively documented for photocatalytic activity in reducing dye pollution are $ZnFe_2O_4$, $MnFe_2O_4$, $CuFe_2O_4$, and $CoFe_2O_4$ [17].

DIFFERENT STRATEGIES FOR EFFICIENT PHOTOCATALYST

To remove pollutants from wastewater, different strategies are utilized to decrease the bandgap, improve semiconducting qualities, and boost photocatalytic activity. Strategies included (i) Using dopants, (ii) Doping with carbon-based materials, and (iii) Using surfactants.

Using of Dopant

The dopant is the foreign substance that is added to the host material to change its properties and provide the desired level of efficiency (Table **1**). Dopants may be added depending on the applications. To improve the optical behaviour of the metal oxides and sulphides, rare-earth metals were additionally added as dopants in addition to Ni, Cu, and Mn [18]. The introduction of the ideal amount of dopant will only assist the material's optical response [19]. Smaller doses of rare earth metal used in rare earth metal doping will cause more pronounced alterations in the host. The rare earth metal dopants had a significant impact on the surface defects, crystal size, bandgap, and structure [20]. Increased photocatalytic effectiveness is made possible by the dopant's ability to suppress electron-hole recombination [21]. Due to its special ability to build complexes over f-orbitals that excitedly tune up the photocatalytic activity, the inclusion of rare earth metals has drawn more attention [22]. The efficiency and dyes utilized for both unaltered and doped photocatalysts are shown in the table below:

Table 1. Efficiency of transition metal and rare earth metal doped photocatalysts.

Sample	Model Pollutant	Light Used	Efficiency (%)	Refs.
Ni- ZnO	Congo Red	UV- Vis	100	[23]
3% Ce- CuO	Methylene Blue	UV	32	[24]
5% Ni doped TiO_3	Rhodamine B	UV	96	[25]
4% Co- α Fe_2O_3	Methylene Blue	UV	92	[26]
2% Sm- $ZnFe_2O_4$	Methylene Blue	Visible	65	[27]
2% Ce- $CuFe_2O_4$	Rhodamine B	Visible	88	[28]
1 M Sn- Co_3O_4	MB	Visible	75	[29]

Using Surfactants

The performance of the photocatalyst under light irradiation may be significantly influenced by its morphology. Numerous synthetic techniques were used to change the morphology [30]. Surfactant, which promotes better growth of various morphologies, is included here. By adsorbing molecules on crystals of nucleation centres, surfactants, which have both hydrophobic and hydrophilic properties, will control the morphologies of photocatalysts. Surfactants have been employed as a structure-directing agent for all materials since they were first used to create single crystals without imperfections. Surfactants are added to photocatalysts, which change not only the forms but also the surface area, numerous pores, and uniform structure. These alterations will alter the photocatalyst's characteristics, making it a more effective candidate for light irradiation and more useful for wastewater management [31]. The shape, size, increase of larger surface area and porosity, and better quantum confinement are all clearly directed by surfactants [32]. The impact of surfactants on enhancing photocatalytic efficiency has been summarized in the literature Table **2**.

Table 2. Comparison of photocatalysts efficiency with surfactants.

Sample	Model Pollutant	Light Used	Efficiency (%)	Refs.
PVP- ZnO	MO and MG	105 W	82.7	[33]
$FeVO_4$ - HTAB	Phenol	Solar light	100	[34]
SDS- (Zn- SnO_2)	RhB	Visible light	90	[35]
PEG- $BiVO_4$	RhB	Visible light	95	[36]
2% PVP- (Mn- CdS)	MB	Visible light	98	[37]
CTAB- Bi_2WO_6	RhB	Visible light	98	[38]

Carbon-based Metal Doping

The carbon-based compounds are highly porous and have a huge surface area. The characteristics of the host materials will be improved by doping carbon-based compounds onto the other materials. A better photocatalytic activity is produced at the interface created by the synergistic interaction of carbon-based material and host material [39]. Carbon products were chosen as prospective replacements for other materials in water-based remediations because of their availability and superior qualities. The special characteristics of carbon materials include their large surface area, which allows for significant pollution absorption, and their good electron-saving ability, which promotes charge separation and prevents e- and h+ recombination [40]. The nitrogen-rich polymeric carbon substance known as "Melon" or g-C_3N_4 was synthesized from repeating triazine. There are many uses for g-C_3N_4, including the production of hydrogen, photocatalysis, solar cells, and water filtration. The g-C_3N_4 source altered the chemical and electrical characteristics [41]. Table **3** contains a comparison of the photocatalysts.

Table 3. Comparison table of photocatalysts with carbon materials with efficiency.

Sample	Model pollutant	Light used	Efficiency %	References
g-C_3N_4/TiO_2	RhB	Visible light	97	[42]
g-C_3N_4/ Ag_3VO_4	Basic Fuschin	Visible light	95	[43]
Ag/$FeWO_4$/g-C_3N_4	RhB	Sunlight	98	[44]
g-C_3N_4/BiOBr	MO	Visible light	94	[45]
MnO_2/g- C_3N_4	MO	Visible light	65	[46]

CONCLUSION

The art of wastewater management involves effectively reusing water. According to experts, photocatalysis is the best process for managing wastewater effectively. In this time period, many different photocatalysts are being researched. There are still numerous research projects being conducted to create a possible catalyst. To improve the effectiveness of degrading organic contaminants, we have reviewed several of the materials with various modifications. The action of the photocatalyst is significantly improved by the inclusion of dopants. Morphology is produced by the addition of surfactants since they improve the activity. This review will be more helpful in facilitating the development of improved photocatalysts.

ACKNOWLEDGEMENT

This work was supported by UGC-SAP, DST-FIST, DST-PURSE, MHRD-RUSA grants.

REFERENCES

[1]   A. Sonune, and R. Ghate, "Developments in wastewater treatment methods", *Desalination,* vol. 167, pp. 55-63, 2004.
       [http://dx.doi.org/10.1016/j.desal.2004.06.113]

[2]   G. Crini, and E. Lichtfouse, "Advantages and disadvantages of techniques used for wastewater treatment", *Environ. Chem. Lett.,* vol. 17, no. 1, pp. 145-155, 2019.
       [http://dx.doi.org/10.1007/s10311-018-0785-9]

[3]   S. Keerthana, R. Yuvakkumar, P.S. Kumar, G. Ravi, D. Velauthapillai, and D.V.N. Vo, "Investigation of EG-Bi2S3 nanorods photocatalytic activity under visible light for dye degradation from aquatic system", *Environ. Sci. Pollut. Res. Int.,* vol. 30, no. 28, pp. 71628-71636, 2021.
       [http://dx.doi.org/10.1007/s11356-021-14933-z] [PMID: 34156624]

[4]   T. Robinson, G. McMullan, R. Marchant, and P. Nigam, "Remediation of dyes in textile effluent: A critical review on current treatment technologies with a proposed alternative", *Bioresour. Technol.,* vol. 77, no. 3, pp. 247-255, 2001.
       [http://dx.doi.org/10.1016/S0960-8524(00)00080-8] [PMID: 11272011]

[5]   N. Azbar, T. Yonar, and K. Kestioglu, "Comparison of various advanced oxidation processes and chemical treatment methods for COD and color removal from a polyester and acetate fiber dyeing effluent", *Chemosphere,* vol. 55, no. 1, pp. 35-43, 2004.
       [http://dx.doi.org/10.1016/j.chemosphere.2003.10.046] [PMID: 14720544]

[6]   H. Wang, C. Xie, W. Zhang, S. Cai, Z. Yang, and Y. Gui, "Comparison of dye degradation efficiency using ZnO powders with various size scales", *J. Hazard. Mater.,* vol. 141, no. 3, pp. 645-652, 2007.
       [http://dx.doi.org/10.1016/j.jhazmat.2006.07.021] [PMID: 16930825]

[7]   V. Katheresan, J. Kansedo, and S.Y. Lau, "Efficiency of various recent wastewater dye removal methods: A review", *J. Environ. Chem. Eng.,* vol. 6, no. 4, pp. 4676-4697, 2018.
       [http://dx.doi.org/10.1016/j.jece.2018.06.060]

[8]   J.M. Poyatos, M.M. Muñio, M.C. Almecija, J.C. Torres, E. Hontoria, and F. Osorio, "Advanced oxidation processes for wastewater treatment: State of the art", *Water Air Soil Pollut.,* vol. 205, no. 1, pp. 187-204, 2009.

[9]   S.C. Ameta, and R. Ameta, *Advanced Oxidation Processes for Wastewater Treatment.* Academic Press, 2018.

[10]   L. Wang, L. Shen, L. Zhu, H. Jin, N. Bing, and L. Wang, "Preparation and photocatalytic properties of SnO$_2$coated on nitrogen-doped carbon nanotubes", *J. Nanomater.,* vol. 2012, pp. 1-6, 2012.
[http://dx.doi.org/10.1155/2012/909473]

[11]   D.B. Miklos, C. Remy, M. Jekel, K.G. Linden, J.E. Drewes, and U. Hübner, "Evaluation of advanced oxidation processes for water and wastewater treatment : A critical review", *Water Res.,* vol. 139, pp. 118-131, 2018.
[http://dx.doi.org/10.1016/j.watres.2018.03.042] [PMID: 29631187]

[12]   R. Fatima, M.F. Warsi, S. Zulfiqar, S.A. Ragab, I. Shakir, and M.I. Sarwar, "Nanocrystalline transition metal oxides and their composites with reduced graphene oxide and carbon nanotubes for photocatalytic applications", *Ceram. Int.,* vol. 46, no. 10, pp. 16480-16492, 2020.
[http://dx.doi.org/10.1016/j.ceramint.2020.03.213]

[13]   S.H. Baeck, T.F. Jaramillo, C. Braendli, and E.W. McFarland, "Combinatorial electrochemical synthesis and characterization of tungsten-based mixed-metal oxides", *J. Comb. Chem.,* vol. 34, no. 10, 2003.
[PMID: 12425600]

[14]   Y. Li, H. Dong, L. Li, L. Tang, R. Tian, R. Li, J. Chen, Q. Xie, Z. Jin, J. Xiao, S. Xiao, and G. Zeng, "Recent advances in waste water treatment through transition metal sulfides-based advanced oxidation processes", *Water Res.,* vol. 192, p. 116850, 2021.
[http://dx.doi.org/10.1016/j.watres.2021.116850] [PMID: 33513467]

[15]   A. Rafiq, M. Imran, M. Aqeel, M. Naz, M. Ikram, and S. Ali, "Study of transition metal ion doped cds nanoparticles for removal of dye from textile wastewater", *J. Inorg. Organomet. Polym. Mater.,* vol. 30, no. 6, pp. 1915-1923, 2020.
[http://dx.doi.org/10.1007/s10904-019-01343-5]

[16]   E. Casbeer, V.K. Sharma, and X.Z. Li, "Synthesis and photocatalytic activity of ferrites under visible light: A review", *Separ. Purif. Tech.,* vol. 87, pp. 1-14, 2012.
[http://dx.doi.org/10.1016/j.seppur.2011.11.034]

[17]   N. Ali, A. Said, F. Ali, F. Raziq, Z. Ali, M. Bilal, L. Reinert, T. Begum, and H.M.N. Iqbal, "Photocatalytic degradation of congo red dye from aqueous environment using cobalt ferrite nanostructures: Development, characterization, and photocatalytic performance", *Water Air Soil Pollut.,* vol. 231, no. 2, p. 50, 2020.
[http://dx.doi.org/10.1007/s11270-020-4410-8]

[18]   M. Junaid, M. Imran, M. Ikram, M. Naz, M. Aqeel, H. Afzal, H. Majeed, and S. Ali, "The study of Fe-doped CdS nanoparticle-assisted photocatalytic degradation of organic dye in wastewater", *Appl. Nanosci.,* vol. 9, no. 8, pp. 1593-1602, 2019.
[http://dx.doi.org/10.1007/s13204-018-0933-3]

[19]   M. Shaban, A.M. Ahmed, N. Shehata, M.A. Betiha, and A.M. Rabie, "Ni-doped and Ni/Cr co-doped TiO$_2$ nanotubes for enhancement of photocatalytic degradation of methylene blue", *J. Colloid Interface Sci.,* vol. 555, pp. 31-41, 2019.
[http://dx.doi.org/10.1016/j.jcis.2019.07.070] [PMID: 31377646]

[20]   P. Pascariu, C. Cojocaru, N. Olaru, P. Samoila, A. Airinei, M. Ignat, L. Sacarescu, and D. Timpu, "Novel rare earth (RE-La, Er, Sm) metal doped ZnO photocatalysts for degradation of Congo-Red dye: Synthesis, characterization and kinetic studies", *J. Environ. Manage.,* vol. 239, pp. 225-234, 2019.
[http://dx.doi.org/10.1016/j.jenvman.2019.03.060] [PMID: 30901700]

[21]   M.A. Moiz, A. Mumtaz, M. Salman, H. Mazhar, M.A. Basit, S.W. Husain, and M. Ramzan, "Enhancement of dye degradation by zinc oxide *via* transition-metal doping: A review", *J. Electron. Mater.,* vol. 50, no. 9, pp. 5106-5121, 2021.

[http://dx.doi.org/10.1007/s11664-021-09093-2]

[22]   U. Alam, A. Khan, D. Ali, D. Bahnemann, and M. Muneer, "Comparative photocatalytic activity of sol–gel derived rare earth metal (La, Nd, Sm and Dy)-doped ZnO photocatalysts for degradation of dyes", *RSC Adv.,* vol. 8, no. 31, pp. 17582-17594, 2018.
[http://dx.doi.org/10.1039/C8RA01638K] [PMID: 35539270]

[23]   S.M. Mousavi, A.R. Mahjoub, and R. Abazari, "Facile green fabrication of nanostructural Ni-doped ZnO hollow sphere as an advanced photocatalytic material for dye degradation", *J. Mol. Liq.,* vol. 242, pp. 512-519, 2017.
[http://dx.doi.org/10.1016/j.molliq.2017.07.050]

[24]   N. Ekthammathat, A. Phuruangrat, T. Thongtem, and S. Thongtem, "Synthesis and characterization of Ce-doped CuO nanostructures and their photocatalytic activities", *Mater. Lett.,* vol. 167, pp. 266-269, 2016.
[http://dx.doi.org/10.1016/j.matlet.2016.01.020]

[25]   H. Khojasteh, M. Salavati-Niasari, and S. Mortazavi-Derazkola, "Synthesis, characterization and photocatalytic properties of nickel-doped TiO$_2$ and nickel titanate nanoparticles", *J. Mater. Sci. Mater. Electron.,* vol. 27, no. 4, pp. 3599-3607, 2016.
[http://dx.doi.org/10.1007/s10854-015-4197-3]

[26]   S. Keerthana, R. Yuvakkumar, G. Ravi, P. Kumar, M.S. Elshikh, H.H. Alkhamis, A.F. Alrefaei, and D. Velauthapillai, "A strategy to enhance the photocatalytic efficiency of α-Fe$_2$O$_3$", *Chemosphere,* vol. 270, p. 129498, 2021.
[http://dx.doi.org/10.1016/j.chemosphere.2020.129498] [PMID: 33422995]

[27]   S.P. Keerthana, R. Yuvakkumar, P.S. Kumar, G. Ravi, and D. Velauthapillai, "Rare earth metal (Sm) doped zinc ferrite (ZnFe$_2$O$_4$) for improved photocatalytic elimination of toxic dye from aquatic system", *Environ. Res.,* vol. 197, p. 111047, 2021.
[http://dx.doi.org/10.1016/j.envres.2021.111047] [PMID: 33781773]

[28]   S. Keerthana, R. Yuvakkumar, G. Ravi, S. Pavithra, M. Thambidurai, C. Dang, and D. Velauthapillai, "Pure and Ce-doped spinel CuFe$_2$O$_4$ photocatalysts for efficient rhodamine B degradation", *Environ. Res.,* vol. 200, p. 111528, 2021.
[http://dx.doi.org/10.1016/j.envres.2021.111528] [PMID: 34139226]

[29]   S.P. Keerthana, R. Yuvakkumar, P.S. Kumar, G. Ravi, D.V.N. Vo, and D. Velauthapillai, "Influence of tin (Sn) doping on Co$_3$O$_4$ for enhanced photocatalytic dye degradation", *Chemosphere,* vol. 277, p. 130325, 2021.
[http://dx.doi.org/10.1016/j.chemosphere.2021.130325] [PMID: 33774254]

[30]   Q. Liang, X. Liu, G. Zeng, Z. Liu, L. Tang, B. Shao, Z. Zeng, W. Zhang, Y. Liu, M. Cheng, W. Tang, and S. Gong, "Surfactant-assisted synthesis of photocatalysts: Mechanism, synthesis, recent advances and environmental application", *Chem. Eng. J.,* vol. 372, pp. 429-451, 2019.
[http://dx.doi.org/10.1016/j.cej.2019.04.168]

[31]   X. Xu, Z. Gao, Z. Cui, Y. Liang, Z. Li, S. Zhu, X. Yang, and J. Ma, "Synthesis of Cu$_2$O Octadecahedron/TiO$_2$ quantum dot heterojunctions with high visible light photocatalytic activity and high stability", *ACS Appl. Mater. Interfaces,* vol. 8, no. 1, pp. 91-101, 2016.
[http://dx.doi.org/10.1021/acsami.5b06536] [PMID: 26651845]

[32]   K. Sujatha, T. Seethalakshmi, A.P. Sudha, and O.L. Shanmugasundaram, "Photocatalytic activity of pure, Zn doped and surfactants assisted Zn doped SnO$_2$ nanoparticles for degradation of cationic dye", *Nano-Structures & Nano-Objects,* vol. 18, p. 100305, 2019.
[http://dx.doi.org/10.1016/j.nanoso.2019.100305]

[33]   S.M. Lam, M.W. Kee, and J.C. Sin, "Influence of PVP surfactant on the morphology and properties of ZnO micro/nanoflowers for dye mixtures and textile wastewater degradation", *Mater. Chem. Phys.,* vol. 212, pp. 35-43, 2018.
[http://dx.doi.org/10.1016/j.matchemphys.2018.03.002]

[34]   B. Ozturk, and G.S. Pozan SOYLU, "Synthesis of surfactant-assisted FeVO₄ nanostructure: Characterization and photocatalytic degradation of phenol", *J. Mol. Catal. Chem.,* vol. 398, pp. 65-71, 2015.
[http://dx.doi.org/10.1016/j.molcata.2014.11.013]

[35]   S. Keerthana, R. Yuvakkumar, G. Ravi, M. Manimegalai, M. Pannipara, A.G. Al-Sehemi, R.A. Gopal, M.M. Hanafiah, and D. Velauthapillai, "Investigation on (Zn) doping and anionic surfactant (SDS) effect on SnO₂ nanostructures for enhanced photocatalytic RhB dye degradation", *Environ. Res.,* vol. 199, p. 111312, 2021.
[http://dx.doi.org/10.1016/j.envres.2021.111312] [PMID: 34019891]

[36]   M. Shang, W. Wang, L. Zhou, S. Sun, and W. Yin, "Nanosized BiVO₄ with high visible-light-induced photocatalytic activity: Ultrasonic-assisted synthesis and protective effect of surfactant", *J. Hazard. Mater.,* vol. 172, no. 1, pp. 338-344, 2009.
[http://dx.doi.org/10.1016/j.jhazmat.2009.07.017] [PMID: 19632047]

[37]   S. Keerthana, R. Yuvakkumar, G. Ravi, A.E.Z.M.A. Mustafa, A.A. Al-Ghamdi, M. Soliman Elshikh, and D. Velauthapillai, "PVP influence on Mn–CdS for efficient photocatalytic activity", *Chemosphere,* vol. 277, p. 130346, 2021.
[http://dx.doi.org/10.1016/j.chemosphere.2021.130346] [PMID: 33780675]

[38]   H. Zheng, W. Guo, S. Li, R. Yin, Q. Wu, X. Feng, N. Ren, and J-S. Chang, "Surfactant (CTAB) assisted flower-like Bi2WO6 through hydrothermal method: Unintentional bromide ion doping and photocatalytic activity", *Catal. Commun.,* vol. 88, pp. 68-72, 2017.
[http://dx.doi.org/10.1016/j.catcom.2016.09.030]

[39]   N.M. Mahmoodi, "Photocatalytic degradation of dyes using carbon nanotube and titania nanoparticle", *Water Air Soil Pollut.,* vol. 224, no. 7, p. 1612, 2013.
[http://dx.doi.org/10.1007/s11270-013-1612-3]

[40]   K. Fan, Q. Chen, J. Zhao, and Y. Liu, "Preparation of MnO₂-carbon materials and their applications in photocatalytic water treatment", *Nanomaterials,* vol. 13, no. 3, p. 541, 2023.
[http://dx.doi.org/10.3390/nano13030541] [PMID: 36770501]

[41]   V. Devthade, D. Kulhari, and S.S. Umare, "Role of precursors on photocatalytic behavior of graphitic carbon nitride", *Mater. Today Proc.,* vol. 5, no. 3, pp. 9203-9210, 2018.
[http://dx.doi.org/10.1016/j.matpr.2017.10.045]

[42]   D. Monga, and S. Basu, "Enhanced photocatalytic degradation of industrial dye by g-C₃N₄/TiO₂ nanocomposite: Role of shape of TiO₂", *Adv. Powder Technol.,* vol. 30, no. 5, pp. 1089-1098, 2019.
[http://dx.doi.org/10.1016/j.apt.2019.03.004]

[43]   S. Wang, D. Li, C. Sun, S. Yang, Y. Guan, and H. He, "Synthesis and characterization of g-C₃N₄/Ag₃VO₄ composites with significantly enhanced visible-light photocatalytic activity for triphenylmethane dye degradation", *Appl. Catal. B,* vol. 144, pp. 885-892, 2014.
[http://dx.doi.org/10.1016/j.apcatb.2013.08.008]

[44]   R. Saher, M.A. Hanif, A. Mansha, H.M.A. Javed, M. Zahid, N. Nadeem, G. Mustafa, A. Shaheen, and O. Riaz, "Sunlight-driven photocatalytic degradation of rhodamine B dye by Ag/FeWO₄/g-C₃N₄ composites, *Int. J. Environ. Sci. Technol.,* vol. 18, no. 4, pp. 927-938, 2021.
[http://dx.doi.org/10.1007/s13762-020-02888-6]

[45]   M. Jiang, Y. Shi, J. Huang, L. Wang, H. She, J. Tong, B. Su, and Q. Wang, "Synthesis of Flowerlike g-C₃N₄ /BiOBr with enhanced visible light photocatalytic activity for dye degradation", *Eur. J. Inorg. Chem.,* vol. 2018, no. 17, pp. 1834-1841, 2018.
[http://dx.doi.org/10.1002/ejic.201800110]

[46]   S. Panimalar, R. Uthrakumar, E.T. Selvi, P. Gomathy, C. Inmozhi, K. Kaviyarasu, and J. Kennedy, "Studies of MnO₂/g-C₃N₄ hetrostructure efficient of visible light photocatalyst for pollutants degradation by sol-gel technique", *Surf. Interfaces,* vol. 20, p. 100512, 2020.
[http://dx.doi.org/10.1016/j.surfin.2020.100512]

CHAPTER 8

Recent Techniques in Dye Degradation: A Biological Approach

Nagaraj Revathi[1], Jeyaraj Dhaveethu Raja[2], Jegathalaprathaban Rajesh[3] and **Murugesan Sankarganesh[3,*]**

[1] *Department of Chemistry, Ramco Institute of Technology, Rajapalayam, Virudhunagar-626 117, Tamil Nadu, India*

[2] *Department of Chemistry, The American College, Tallakkulam, Madurai-625 002, Tamil Nadu, India*

[3] *Department of Chemistry, Saveetha School of Engineering, Saveetha Institute of Medical and Technical Sciences, Saveetha University, Chennai, Tamil Nadu-602 105, India*

Abstract: Synthetic dyes are organic compounds that are mostly employed in the manufacturing industry. A huge number of dyes are unbound and released into the environment during the dying process. The discharge of dye/effluent with a high biological oxygen demand (BOD) and chemical oxygen demand (COD) into the environment has several negative consequences for the area's flora and fauna. They are poisonous and mutagenic, and have other significant negative impacts on a variety of creatures, including unicellular and multicellular organisms. Besides the costly Physico-chemical treatment methods, biological approaches involving bacteria, fungi, algae, plants, and their enzymes have got a lot of attention in recent years for the decolorization and degradation of dyes contained in effluents due to their economic effectiveness and environmental friendliness. Microbial degradation appears to be the most promising of these technologies for resource recovery and sustainability. Microorganism and plant-derived enzymes' ability to decolorize and break down dyes has long been known, and they are shown to be the most effective molecular weapon for bioremediation. Several sophisticated approaches are currently being investigated for the effective decolorization of textile dyes as well as eco-toxic effluent, including genetic engineering, nanotechnology, mobilized cells or enzymes, biofilms, and microbial fuel cells, among others. These biological methods for decolorization and degradation of textile effluent are very successful and have various advantages over traditional procedures. Biological methods for removing toxic textile dyes are both environmentally friendly and cost-effective.

Keywords: Dye degradation, Dye contaminated industrial effluents treatment, Graphene oxide, Nanocomposites.

---

*Corresponding author **Murugesan Sankarganesh**: Department of Chemistry, Saveetha School of Engineering, Saveetha Institute of Medical and Technical Sciences, Saveetha University, Chennai, Tamil Nadu-602 105, India; E-mail: msankarajan1990@gmail.com

Paulpandian Muthu Mareeswaran & Jegathalaprathaban Rajesh (Eds.)
All rights reserved-© 2023 Bentham Science Publishers

INTRODUCTION

Colour has always fascinated humankind, for both aesthetic and social reasons. A dye is a colored substance that is typically used in solution and can be fixed to the fabric. It is necessary for the dye to be "fast" or chemically stable so that it does not wash out with soap and water.

The most popular industrial coloring chemical substances are dyes. Dyes are organic substances that have a distinct color. Contrary to most organic substances, dyes have color because they:

1. Absorb visible spectrum light (400-700 nm).

2. Have a chromophore (color bearing group).

3. Have a conjugated system.

4. Display electron resonance.

The color is lost when one or more of these characteristics is missing from the molecular structure. Most dyes also contain groups referred to as auxochromes (color aids), such as carboxylic acids, sulfonic acids, amino and hydroxyl groups, in addition to chromophores. Although they do not produce colour, they are frequently utilized to affect dye solubility and their presence can change the color of a colorant. The correlations between visible wavelength and colour absorbed/observed are shown in Table **1**. Almost all commercial products require colour at some point in their production and there are currently more than 50000 trade names for more than 9000 different colourants. The huge quantity results from the wide variety of tints and colours needed, the chemical makeup of the materials to be coloured and the fact that colour is inversely proportional to dye's molecular structure.

Table 1. **Wavelength of light absorption versus colour in organic dyes.**

S. No.	Wavelength Absorbed (nm)	Colour Absorbed	Colour Observed
1.	400-435	Violet	Yellow-Green
2.	435-480	Blue	Yellow
3.	480-490	Green-Blue	Orange
4.	490-500	Blue-Green	Red
5.	500-560	Green	Purple
6.	560-580	Yellow-Green	Violet
7.	580-595	Yellow	Blue

(Table 1) cont.....

S. No.	Wavelength Absorbed (nm)	Colour Absorbed	Colour Observed
8.	595-605	Orange	Green-Blue
9.	605-700	Red	Blue-Green

The textile industry in India accounts for 14% of all industrial production, 4% of the country's GDP and around 27% of all foreign exchange earnings. Up to 10,000 dyes are accessible worldwide, and they have produced more than 7,105 tonnes annually. These colours are employed not only in the textile industry but also in the paper, food and pharmaceutical sectors. The processing of 1 kg of textiles in India requires more than 100 L of water and the textile industry has significantly impacted the contamination of ground and surface water supplies over most of the nation.

Due to its extensive use of up to 8000 chemicals and large amounts of water, the textile industry has been one of the main pollutants of surface and groundwater resources. According to some sources, an average-sized textile sector uses 1.6 million liters of water per day to produce 8000 kg of cloth [1]. The World Bank estimates that the dyeing and finishing processes used on the cloth account for 17% to 20% of the water pollution caused by the textile sector [2, 3]. Thirty of the 72 hazardous compounds found in the wastewater released from the textile dyeing industry cannot be eliminated by wastewater treatment procedures [4]. The textile and dyeing industries are using chemical-based textile dyes more and more because they are more affordable and have higher temperature and light stability than natural colours.

Dye molecules are comprised often key components:

• Chromophores: They are responsible for producing the color. Fig. (**1**) represents various chromophores present in organic dyes.

• Auxochromes: They can not only increase the chromophore content but also make the molecule water soluble and increase its affinity for the fiber [5].

EFFECT OF DYES

The textile industry and its wastewater have grown proportionally along with the demand for textile products, making it one of the major contributors to the world's serious pollution issues. The colouring (textile, cosmetic and leather) industries utilize around 1,00,000 commercial dyes and dyestuff and roughly 10-15% of all dyestuff is lost to wastewater directly. The textile dye industry involves various processes like printing, dyeing, mercerizing, bleaching, scouring, desizing for the production of fabrics, which release textile effluents (Fig. **2**). The effluents are released directly into natural water bodies. These dyes are stable and can linger in the environment for a long time without proper treatment. As a result, this effluent needs to be treated before being released into freshwater streams. Untreated textile effluent's harmful and carcinogenic effects are well known. The dyeing and finishing of textiles produce a lot of wastewater that contains dye which is one of the main causes of water pollution issues worldwide.

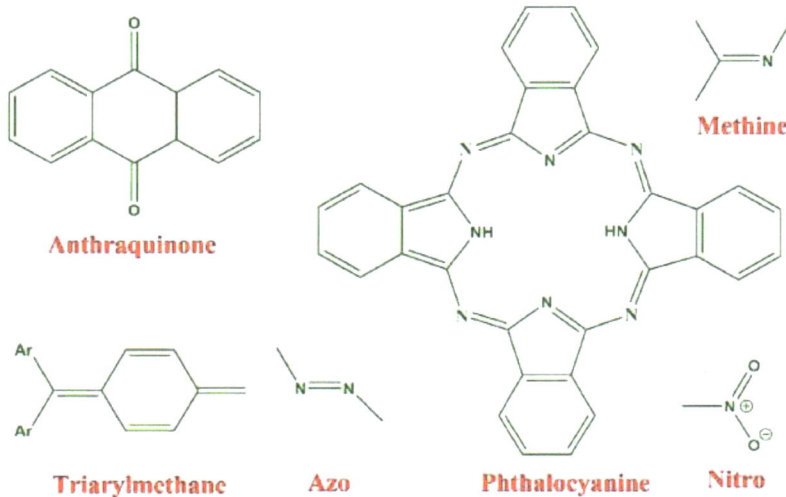

Fig. (1). Chromophores present in organic dyes.

Although the concentration of dyes in wastewater is typically lower than the other chemicals present (less than 1 ppm for some dyes), they frequently attract the most attention due to their strong colour, which renders them highly visible even at very low concentrations, causing serious aesthetic and pollution problems in wastewater disposal as well as water transparency and gas solubility in lakes, rivers and other water bodies [6]. The removal of soluble colourless organic substances, which often make up the majority of BOD, is frequently less crucial than the removal of colour from wastewater [7]. In the colour index, there are over 8000 chemical products related to the dyeing process listed and more than 1,00,000 commercial dyes and dyestuffs are used in the colouring (textile, cosmetic and leather) industries [8, 9]. Additionally, there are over 7×10^5 metric tonnes of dyestuff produced annually [10, 11]. 10-15% of the dyes used in the industrial dyeing process are lost in the textile factories' effluents, giving their products a highly colourful appearance [12, 13]. In such industrial effluents around the world, 2,80,000 tonnes of textile dyes are thought to be released annually [14]. Under anaerobic conditions, direct discharge of these effluents results in hazardous aromatic amine production in receiving media. Water pollution control is today one of the main topics of scientific work because many synthetic dyes are poisonous, mutagenic, and carcinogenic [15], in addition to their negative effects on COD and their cosmetic effect.

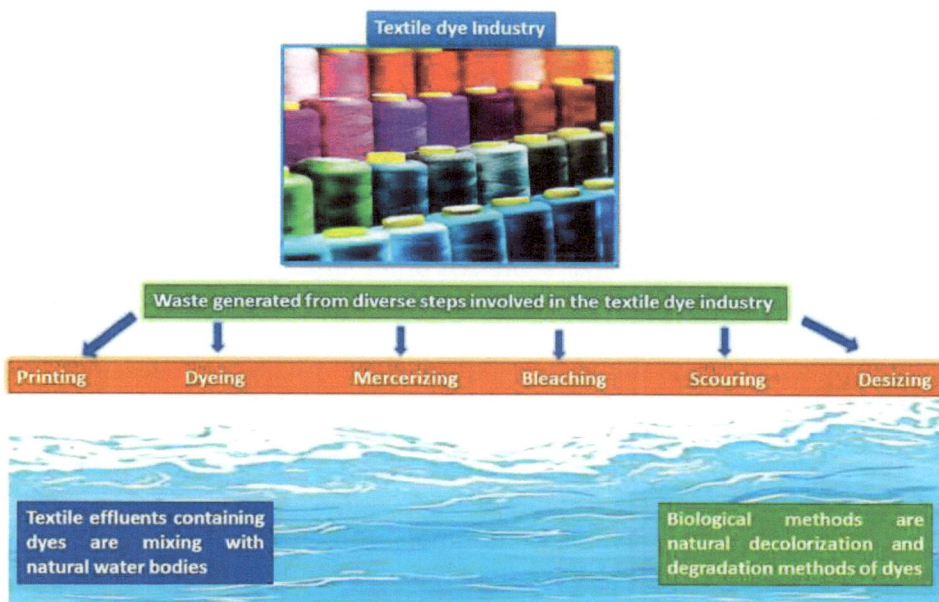

Fig. (2). Effect of dyes by various processes.

The quest for suitable treatment methods is a top concern due to the typically high volumetric rate of industrial wastewater discharge and increasingly strict legislation [16]. Dye-containing textile industry effluents are vividly coloured and easily recognizable [17]. Traditional wastewater treatment is still inefficient due to the dyes' complex aromatic structure, which is resistant to light, biological activity, ozone and other deteriorating environmental factors [18]. Additionally, due to the reactive cleavage of azo groups, anionic and non-ionic azo dyes emit hazardous amines [19]. Another environmental problem is the presence of heavy metals, including chromium, cobalt and copper, in wastewater [20]. It has been a major difficulty for scientists up to this point to come up with a simple, cost-effective technology for treating dyes in textile wastewater [18, 21].

TREATMENT METHODS

Chemical and physical treatments are the most often used techniques for the decolorization and degradation of dyes, although most of these techniques have drawbacks, including high operating costs and disposal of a lot of sludge generated during these procedures. As a result, most studies have concentrated on adopting the least expensive and most environmentally friendly techniques, such as biological processes. Systems for treating industrial biological wastewater use microorganisms to eliminate contaminants from the environment. The decomposition of the organic materials is brought on by the microorganisms employed. Biological treatments provide many benefits, including being affordable, straightforward, producing less surplus sludge and being highly flexible because they may be used with a wide variety of effluents. One of the most significant sectors of the global economy and the fifth-largest supplier of foreign exchange, the textile industry is also one of the major contributors to water pollution.

Most research findings emphasize *white rot* fungi and their degradative enzymes. Therefore, the key difficulty is not only to find a local, efficient strain that has a high percentage of decolorization capacity for wastewater but also to identify the mechanism by which these fungi carry out the decolorization process. Logrono *et al.,* examined advanced Physico-chemical techniques that integrated biological techniques to lower chemical oxygen demand by up to 98% and remove heavy metals like chromium (54-80%) and zinc (98%) [22]. Trucanu *et al.* [23] did a second investigation on the cathodic decolorization of reactive dyes from the textile industry. By using cyclic voltammetry on a suspended mercury drop electrode, they discovered that the breakdown of chromophores in azo-bonded dyes was caused by the transport of electrons from the cathode. Rajasimman *et al.* also investigated biosorption for the treatment of textile effluent using an anaerobic sequential batch reactor with a decolorization efficiency of up to 94.8% utilizing low-cost adsorbents such as pulverized shell powder [24].

In a triple-layered fixed bed reactor, Kurade *et al.*, demonstrated that the yeast consortium performed more effectively than individual organisms at decolorizing textile effluent [25]. Textile effluent is produced as a result of the numerous fiber processing stages used in the textile industry. To effectively treat textile wastewater, researchers are currently working to develop and apply new treatment solutions, such as hybrid technology, a combination of physical and biological techniques. Several physicochemical decolorization processes have been published during the past 20 years, but only a small number have been adopted by the textile industries [26, 27]. To decolorize dyes in effluents, companies and scientists are under increasing pressure to develop novel, innovative treatments, and technologies because many textile facilities are located in rural areas and municipal treatment expenses are high [28].

The most crucial part of complying with environmental standards is the decolorization and detoxification of industrial dye effluents. There is an urgent need to limit the introduction of azo dyes into the environment due to their negative impact on human health and the ecosystem. However, it is not realistically possible. Adopting treatments that lessen or eliminate colours from wastewater is thus a solution to this issue. Numerous Physico-chemical techniques have been used to treat textile wastewater, but these techniques have several drawbacks because of their high cost, poor efficiency, and issues with secondary pollution. Because they are more affordable and environmentally benign than physical-chemical procedures, biological approaches—which include bacteria, fungi, yeast, algae, plants and their enzymes—have gained popularity. Both biosorption and biodegradation are biological processes that can decolorize azo dyes. The breakdown of dyes may also involve several reductive and oxidative enzymes.

The main area of concern nowadays is finding affordable and efficient ways to handle wastewater discharged by the textile industries to safeguard the aquatic life in water bodies. To provide effective solutions in eliminating contaminants from wastewater originating from the textile industry, the approaches could be physico-chemical, biochemical, or a mix of the two. Williams stated that while physical and chemical approaches can be economical, they cannot ensure that all impurities would be completely removed.

There are two primary categories for the decolorization of pigments in wastewater:

• The first step in determining the mechanism is to modify the environment such that the fungus can decolorize dye effluent as much as possible.

• The second and perhaps most practical option is to try to reuse dyes by decolorizing textile effluents.

The following steps are used in biological treatment to remove colours from wastewater.

1. Mycobiota isolation from effluents.

2. The isolation and identification of the mycobiota.

3. Testing the isolated fungi for their ability to remove textile dye colour.

4. Modify co-metabolic conditions to get the greatest decolorization.

5. Calculate the chemical parameters before and following a fungus treatment.

6. Utilizing bacteria to assess for toxicity.

7. The method of action for the detection of dye wastewater decolorization.

8. Make wastewater effluents colourless.

9. Reusing trapped dyes to colour fabric.

Biological Treatments

The formation of concentrated sludge poses a disposal issue because the Physico-chemical methods used to remove colours from effluents are frequently highly expensive. Finding alternative methods that are efficient in removing colours from huge amounts of effluents, affordable and technically appealing is thus necessary [29]. The suggested approach is to use inexpensive, easy-to-use biological technologies. Microorganisms' capacity for dye decolorization has drawn a lot of interest and is regarded as a practical way to remove harmful contaminants from the environment. Recent foundational research has shown that numerous microorganisms exist that can decolorize a wide range of colours [10]. Growing interest has been shown in microbial decolorization using appropriate bacteria, algae and fungi [30]. These microorganisms can biodegrade and/or bioabsorb colours in wastewater [8]. The first report of *Bacillus subtilis* [31], which was followed by reports of *Aeromonas hydrophila* [32], *Bacillus cereus* [33], *Pseudomonas* strains, *Proteus mirabilis* and *Mycobacterium avium*, signaled the beginning of attempts to isolate bacterial cultures capable of degrading azo dyes.

The ability of some *actinomycete* strains to decolorize reactive dyes, including anthraquinone, phthalocyanine and azo groups, has been documented. This is accomplished by the dyes adhering to the cellular biomass without being broken down. The same *actinomycete* cultures completely decolored other Cu-based azo dyes by degradation, including formazan-copper complex colours [34]. It has been proposed to use algae to take colour out of textile wastewater. As suitable biomaterials for the biological treatment of simulated synthetic azo dye (reactive yellow 22) effluents, *Spirogyra* green algae were studied for their possible usage. Their capacity to remove colour was based on both the dye concentration and algal biomass. A strain of yeast called *Candida zeylanoides* destroyed some simple azo dyes in liquid-aerated batch cultures. Only a few investigations on yeast decolorization have been published; these studies examined *Kluyveromycsemarxianus* IMB3's potential decolorization of remazol black B dye.

There has been a great deal of research done on the function of fungus in wastewater treatment. In the treatment of textile effluents and the eradication of dyes, fungi have proven to be an effective organism. Because fungi have a higher cell-to-surface ratio and more enzymatic and physical contact with the environment than single-cell organisms, they have an additional advantage over them in solubilizing the insoluble substrates. The extracellular nature of the fungal enzymes helps them tolerate toxicants in high quantities. Adsorption on microbial biomass (biosorption) or biodegradation of the dyes by live cells are the two processes that can decolorize and degrade azo dyes. Due to its capacity to break down practically all dye materials and overcome many Physico-chemical procedures' drawbacks, biological treatment of textile azo dyes is the best way. Numerous research that focuses on the employment of microorganisms to degrade colours indicate that biodegradation is an economical and environmentally friendly way to treat wastewater that contains dyes. Through anaerobic, aerobic and sequential anaerobic-aerobic treatment techniques, it is possible to successfully remove the colour of a variety of azo dyes using microorganisms (bacteria, fungi, algae and yeast), plants and their enzymes.

Numerous fungus genera, both in living and dead forms, have been used to decolorize dyes. Biosorption, bioaccumulation and/or biodegradation are used depending on the mechanism involved in biodegradation. Fig. (3) represents various biological processes like binding process, accumulation process and energy efficient process through biosorption, bioaccumulation and biodegrdation, respectively. Through the action of numerous enzymes, biodegradation is an energy-dependent process that breaks down dye into a variety of byproducts. When using living biomass, biosorption is the simultaneous occurrence of mechanisms that do not require metabolic energy or transport to bind solutes to the biomass. As a result, it can happen in both live and dead biomasses. The buildup of contaminants by metabolically active, actively developing cells is known as bioaccumulation [35].

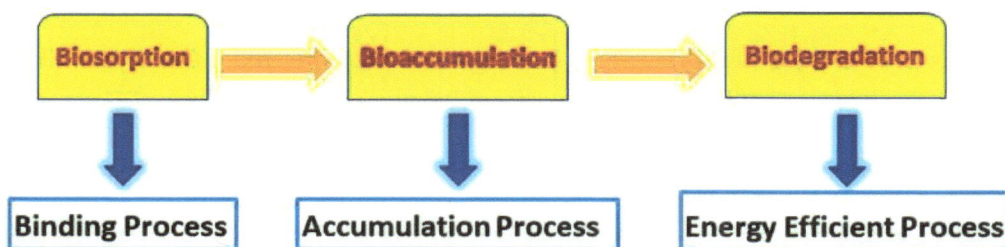

Fig. (3). Various biological treatment processes.

To get microbiological agents suited for decolorizing wastewater containing dye, enrichment processes were developed. Different biological species like algae, bacteria, fungi, yeast, plant and actinomycetes acted as bioagents for dye degradation for the treatment of textile effluents (Fig. **4**). The results of these procedures led to the identification of various fungus strains with decolorization abilities. These include the *Myrothecium verrucaria* and *Ganoderma sp.* strains, which are efficient for a variety of colours, mostly through adsorption to the fungal *mycelium*. It was noted that minimal colour loss occurred at high dye concentrations. In the dye decolorization process by *Phanerochae techrysosporium*, Spadaro *et al.* found that aromatic rings having substituents as hydroxyl, amino, acetamido or nitro functionalities were mineralized to a larger extent than unsubstituted rings [36]. The sensitivity of five sulfonated azo dyes to degradation by *Phanerochaetechrysosporium*, on the other hand, was not significantly influenced by the substitution pattern, according to Paszczynski *et al.* findings' [37].

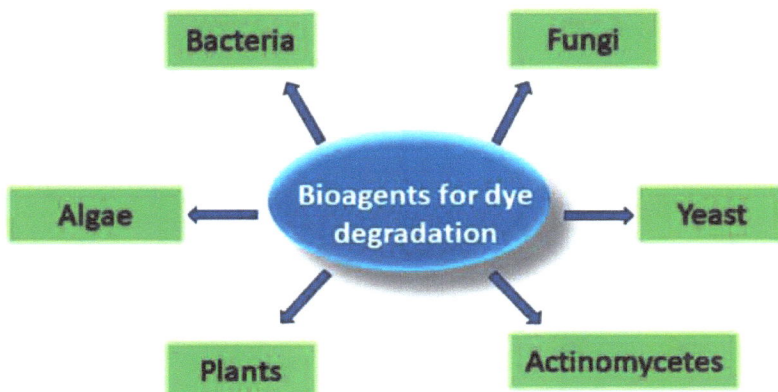

Fig. (4). List of bioagents for dye degradation.

After 3 to 5 days of incubation, an *Aspergillus sojae*B-10 strain was found to be capable of decolorizing the azo dyes amaranth, congo red and Sudan III in nitrogen-poor conditions. According to Wong and Yu, the structure of the dye was important for Trametes Versicolor's ability to decolorize it. Anthraquinone dyes were substrates for laccase, while azo and indigo dyes were not the same enzyme substrates. Additionally, Zhang *et al.,* noted that as cotton bleaching effluent concentration increased, the efficiency of colour removal reduced [38]. *Trichophytonrubrum* LSK-27 has been shown by Yesiladali *et al.,* to be a viable culture for dye removal applications and to be a suitable candidate for the aerobic treatment of textile effluents leading to the harmless breakdown of dye compounds [11].

For dye degradation, practical and environmentally favorable methods are required. Since the initial report of the degradation of dyes by *white rot* fungi in 1983, numerous *white rot* fungi have been investigated for their capacity to decolorize. A variety of colours can be broken down by the non-specific ligninolytic enzyme system found in *white rot* fungi, which also includes laccase, manganese peroxidase and lignin peroxidase. To start the oxidation of substrates in the extracellular environment of the fungal cells, the fungus excretes these ligninolytic enzymes extracellularly. Ohmomo *et al.,* found many fungal cultures that could decolorize things and they named them *Coriolus Versicolor, Mycelia sterilia* and *Aspergillus fumigatus* [39]. In addition, several additional *brown rot* fungi that could decolorize a variety of dyes with various structural differences were discovered to be more efficient than *Phanerochaete chrysosporium* [40].

Additionally, numerous fungi can decolorize and/or biosorb different dyes, including *Aspergillus niger, Rhizopusarrhizus* and *Rhizopusoryzae. Aspergillus niger* is a common saprophyte that grows on dead leaves, stored grain, compost piles and other decaying vegetation. *Aspergillus* species are a widespread genus of filamentous fungus that is frequently isolated from soil, plant detritus and indoor air settings. The spores are widely distributed and frequently found in soil and organic materials. Although perfect forms of fungi that (reproduce sexually) have been discovered, *Aspergillus niger* includes a group of fungi that are typically thought to as asexual. On a variety of artificial surfaces, *Aspergillus niger* develops quickly and produces colonies with septate, *hyaline mycelial hyphae* that are covered in a dense covering of dark brown to black conidial heads. The brewing and textile industries employ a lot of *Aspergillus* enzymes. According to several studies, *Aspergillus* fungi can be utilized to eliminate radioactive and harmful metals from the environment.

The Food and Drug Administration (FDA) considers *Aspergillus niger* fermentation to be generally recognized as safe (GRAS). *Aspergillus niger* is frequently used to test the effectiveness of preservative treatments. Additionally, *Aspergillus niger* strains are used for soil testing because of research demonstrating that it is very sensitive to nutritional deficits. One of the most significant enzymes produced by *Aspergillus niger*, which also produces a variety of other enzymes, is glucose oxidase. A dimeric flavoprotein called glucose oxidase (also known as -D-glucose: oxygen 1-oxidoreductase) catalyzes the conversion of -D-glucose to D-gluconic acid and hydrogen peroxide by oxidizing it with molecular oxygen (as an electron acceptor). A glycosylated enzyme with a total molecular weight of 160.000 daltons and two identical polypeptide chain subunits that are covalently connected by disulfide bonds, glucose oxidase is a member of the broad class of enzymes known as oxidoreductases or "redox enzymes." Co-factors must either be generated or added to the reaction mixture to carry out many different sorts of redox reactions. The electron-transport chains in living systems are used to do this. There are just a few examples of redox enzymes being used in the food industry. One of these is glucose oxidase, which has a co-factor that is a tightly attached prosthetic group that can regenerate itself by reducing oxygen to hydrogen peroxide.

The carbon source is just one of many factors that are crucial to dye decolorization. According to Kapdan *et al.,* glucose is the most easily utilizable carbon source for most fungi [41]. In the decolorization of everzol turquoise blue g by *Coriolus Versicolor*, they used starch, molasses, fructose and glucose and they saw that glucose showed the best results, in which case 100% dye removal was obtained after 5 days of incubation, followed by fructose 92% in 9 days. Additionally, the amount of nitrogen in the media impacts dye decolorization by changing the fungi's ability to produce certain enzymes. Three nitrogen sources (ammonium chloride, peptone, and malt extract) were investigated by Hatvani and Mecs to see how they would affect the decolorization of Poly R-478 by Lentinusedodes. They found that NH_4Cl, peptone and malt extract decolorized the dye under test in 18, 21 and 17 days, respectively. Additionally, it was discovered that textile wastewater contains a variety of acids, alkalis, salts, or metal ions as impurities; the presence of these ions affected the fungus' ability to decolorize dyes [18].

The inclusion of Cr(VI) in the growing medium considerably reduced the growth of the fungus and lengthened the growth period, delaying and slowing both dye removal and removal rate, according to Aksu *et al.*'s report on *Trametes Versicolor* removal of remazol black b [35]. The research revealed that the fungal biomass can remove colours at higher temperatures which is a necessity for fungi's potential usefulness in cleaning dye wastewater and the temperature was therefore regarded to be a crucial element [18]. In addition, Zeroual *et al.* showed that the decolorization of bromophenol blue dye on *Rhizopusstolonifer* biomass rose marginally with temperature rise to 35°C before stabilizing at 55°C [42]. Due to the enhanced surface activity of each dye molecule, Bayramoglu and Arica discovered that *Trametes Versicolor*'s ability for decolorization improved when the temperature was raised from 5 to 35°C for direct blue-1 and direct red-128 [43].

Additionally, pH strongly affects the chemistry of both dye molecules and fungal biomass with the initial pH of the dye solution having a major impact. The degree to which charged dye groups are adsorbed to the adsorbent depends on its surface charge which is affected by the pH of the solution [18]. *Coriolus Versicolor*'s best growth pH was observed by Kapdan *et al.*, to be 4.5; this pH also produced the maximum decolorization effectiveness (99%), which decreased to 50% at pH 6 and 7 [41]. According to O'Mahony *et al.*, the greater protonation of the weak base groups at lower pH accounts for the higher decolorization ability of *Rhizopusarrhizus* for acidic dye remazol blue at pH 2 as compared to that at pH

10 [44]. Following the acquisition of a net positive charge, these base groups bind the anionic groups of the acidic dyes. The sorption of methylene blue by *Fomes fomentarius* and *Phellinus igniarius* increased from 18 to 75% and 16 to 79%, respectively, as the pH was raised from 3 to 11 according to Maurya *et al.* [45]. This is because deprotonation of various functional groups has increased the net electronegativity of the biosorbent which in turn has increased the electrostatic interaction between the negatively charged biosorbent and the positively charged methylene blue ions.

A green way to remove colour from textile effluent with little expense and maximum up-time is biological procedures which involve the degradation of dyes using biological phenomena like bioremediation. Ali proposed using biological materials that can break down and absorb various synthetic colours, such as algae, bacteria, fungi and yeasts [46]. For the successful degradation of the effluent from the textile industries, biologically based approaches have been applied. When compared to other treatments, biological degradation (also known as bioremediation) is more environmentally friendly, economically viable and produces less sludge overall. Because of the bond breaking (*i.e.*, chromophoric group), it facilitates the degradation of synthetic dyes into a somewhat less hazardous inorganic chemical and ultimately aids in colour removal. The breakdown of the azo bonds that create the amines occurs in two phases,

• The first step is the degradation of azo dyes:

• Second, in an aerobic environment, the aromatic amines are further catabolized to produce tiny harmless compounds.

The methods being researched make use of bacteria's capacity to endure both aerobic and anaerobic conditions to completely degrade the azo linkages created inside the dyes. Lewinsky, Lin, and Lo provided the understanding that biological processes are effective at lowering COD and turbidity but ineffective at removing colour. In the development of biological approaches for decolorization in the future, Muda *et al.* reported the success and usefulness of a two-phase process where the first phase involves anaerobic activities and the second phase involves aerobic processes.

Biosorption

Biosorption is the word used to describe the uptake or accumulation of substances by microbial mass. Toxic dyes have been successfully removed by biosorption using biomass from bacteria, yeast, filamentous fungi and algae. This characteristic of microorganisms is caused by the heteropolysaccharides and lipids that make up their cell walls. These compounds contain a variety of functional groups, including amino, hydroxyl, carboxyl, phosphate and other charged groups. These functional groups bind to the azo dye with strong attractive forces. The main benefits of the biosorption process include its great selectivity, efficiency, economy, ability to operate at low concentrations and good removal performance. It has also been demonstrated to be more effective than currently used Physico-chemical techniques like ion exchange.

Fungi

Fungi have also been employed as latent sorbents for the decolorization of azo dyes from industrial effluents and have received significant attention, like bacteria-mediated biosorption. Many factors, including pH, temperature, ionic strength, contact time, adsorbent, dye concentration, dye structure and the kind of microorganisms used, affect how well biosorption works. For cleaning the effluent outflow from dying industries, various fungus species have been identified and tested. For the biological breakdown of textile colours, diverse microorganisms' combinations of aerobic and anaerobic treatment have produced encouraging results. To replace the current chemical and physical treatment procedures, numerous investigations on fungal-based dye decolorization have been done. An effective remediation model for treating textile, pulp and paper industry effluent that contains polycyclic aromatic hydrocarbons is *phanerochaete chrysosporium* (PAH). Senthilkumar *et al.,* investigated *Phanerocheate chrysosporium*, which generates extracellular enzymes for the decolorization of different colours, including lignin peroxidase, manganese peroxidase and laccase [47].

Algae

Due to their availability in both fresh and saline water, algae are also regarded as powerful biosorbents in addition to the use of bacterial and fungi biomass in the biosorption of azo dyes. According to Donmez and Aksu's findings, algae have a high binding affinity and a relatively high surface area, which contribute to their potential for biosorption [48]. Algal biosorption is mediated by electrostatic attraction and complexation in the cell wall of algae. Because they do not require nutrients, can be stored and utilized for extended periods and can be renewed using organic solvents or surfactants, dead algal cells have shown to be more useful as biosorbents than living ones.

Enzymatic Decolorization and Degradation

In comparison to traditional physical-chemical treatments, the enzymatic technique offers an alternate strategy for the decolorization/degradation of azo dyes from wastewater since it produces less sludge and is more affordable. Some enzymes are involved in the removal of azo dyes and these enzymes have proven to be powerful molecular weapons for the decolorization of azo dyes. Reductive and oxidative enzymes are two major categories that are used to categorize enzymes that cause azo dye decolorization.

Reductive and Oxidative Enzymes

The catalytic proteins, known as azoreductases, are produced by microorganisms like bacteria, algae, and yeast [49]. Several bacterial species have been identified that have the capacity of degrading azo dyes under decreased circumstances. Numerous oxidative enzymes, including polyphenol oxidases (PPO), manganese peroxidase (MnP), lignin peroxidase (LiP), laccase (Lac), tyrosinase (Tyr), N-demethylase, dye decolorizing peroxidases and cellobiose dehydrogenase, are also encoded by microorganisms. Bacteria, filamentous fungi, yeast and plants have all been found to contain these oxidases. These enzymes accelerate the transformation of a wide variety of substrates into less harmful insoluble molecules. By a procedure involving the synthesis of free radicals and an insoluble product, hazardous chemicals are removed from waste. Heme-containing peroxidase is commonly present in plants, microbes and animals. For the treatment of wastewater containing coloured contaminants, extensive research has been done in the last several years to develop systems that rely on peroxidases from plants and fungi.

In the oxidation of Methylene Blue (Basic Blue 9) and Azure B dyes, Ferreira-Leitao *et al.* (2007) reported the involvement of plant Horseradish peroxidase and LiP from *Penicillium chrysosporium* [50]. Versatile Peroxidase (VP), isolated from *P. Chrysosporium* along with LiP and MnP, has recently been purified and described as a new family of ligninolytic peroxidases. Interestingly, despite the presence of Mn^{2+}, these enzymes have demonstrated the activity of both LiP and MnP and the ability to oxidize Mn^{2+} to Mn^{3+} at a pH of about 5.0, while aromatic compounds at pH 3.0. Due to their unique characteristics, such as their nonspecific oxidation capacity, lack of dependence on cofactors and inability to utilize easily accessible oxygen as an electron acceptor, laccases, which are multi-copper oxidases, are the most researched enzymes that aid in the removal of dyes. Laccases use redox media (such as ABTS) to speed up the reaction as they catalyze the decolorization of textile dyes through either direct oxidation or indirect oxidation. In the catalytic cycle of MnP, manganous ions (Mn^{2+}) are converted to Mn^{3+}, which is then chelated with organic acids (*e.g.*, oxalic acid). The secondary substrates are oxidized by the enzyme once the chelated Mn^{3+} rapidly diffuses from the enzyme's active site. The tetramer enzyme polyphenol oxidase removes aromatic pollutants from diverse polluted areas by attaching to two aromatic compounds and using oxygen to do so. It has four copper atoms per molecule and two aromatic compound binding sites. O-hydroxylation of monophenols to o-diphenol is catalyzed by PPO. They are also capable of facilitating the conversion of o-diphenols to o-quinones.

CONCLUDING REMARKS

Based on the presence of non-decomposable colours and a variety of harmful elements in the wastewater stream, the wastewater discharge from textile companies to natural water bodies (such as natural ponds, lakes, streams, creeks, and rivers) can be classified. To complete their manufacturing procedures for fabric pretreatment, several textile businesses use a variety of colours and chemicals. Every pretreatment process satisfies the necessary industrial objectives for the sector, but it also produces dangerous waste products that are sometimes released into water bodies untreated. The capacity for degradation depends on various physical characteristics, including the concentration and kind of dye, pH, salinity, and generation of the final product that may be harmful. About specific bacterial cells or an enzyme used for initial treatment to remove colour from the effluent released after the dye bath and rinse procedures, the anaerobic granular sludge and other redox mediators demonstrated effective results that have been well appreciated. Additionally, encouraging outcomes were observed when redox mediators and thermophilic therapy were used to catalyze the decolorization process in the bioreactor. Extracellular glucose oxidase was not shown to be the primary cause of the decolorization of the dyes when enzymatic action was examined *in vitro* utilizing extracellular fluids (ECF). This confirms that the decolorization process by fungal biomass occurs through microbial metabolism and is connected to microbial development. Adsorption was investigated and demonstrated minor decolorization.

REFERENCES

[1]   R.V. Khandare, and S.P. Govindwar, "Phytoremediation of textile dyes and effluents: Current scenario and future prospects", *Biotechnol. Adv.,* vol. 33, no. 8, pp. 1697-1714, 2015.
[http://dx.doi.org/10.1016/j.biotechadv.2015.09.003] [PMID: 26386310]

[2]   C.R. Holkar, A.J. Jadhav, D.V. Pinjari, N.M. Mahamuni, and A.B. Pandit, "A critical review on textile wastewater treatments: Possible approaches", *J. Environ. Manage.,* vol. 182, pp. 351-366, 2016.
[http://dx.doi.org/10.1016/j.jenvman.2016.07.090] [PMID: 27497312]

[3]   B. Rani, R. Maheshwari, R.K. Yadav, D. Pareek, and A. Sharma, "Resolution to provide safe drinking water for sustainability of future perspectives", *Res. J. Chem. Environ. Sci.,* vol. 1, pp. 50-54, 2013.

[4]   H.L. Chen, and L.D. Burns, "Environmental analysis of textile products. Clothing", *Cloth. Text. Res. J.,* vol. 24, no. 3, pp. 248-261, 2006.
[http://dx.doi.org/10.1177/0887302X06293065]

[5]   R.M. Christie, *Colour Chemistry.* The Royal Society of Chemistry: Cambridge, United Kingdom, 2001.
[http://dx.doi.org/10.1039/9781847550590]

[6]   D.E. Kritikos, N.P. Xekoukoulotakis, E. Psillakis, and D. Mantzavinos, "Photocatalytic degradation of reactive black 5 in aqueous solutions: Effect of operating conditions and coupling with ultrasound irradiation", *Water Res.,* vol. 41, no. 10, pp. 2236-2246, 2007.
[http://dx.doi.org/10.1016/j.watres.2007.01.048] [PMID: 17353027]

[7]   I.M. Banat, P. Nigam, D. Singh, and R. Marchant, "Microbial decolorization of textile-dyecontaining effluents: A review", *Bioresour. Technol.,* vol. 58, no. 3, pp. 217-227, 1996.
[http://dx.doi.org/10.1016/S0960-8524(96)00113-7]

[8]   Y. Fu, and T. Viraraghavan, "Fungal decolorization of dye wastewaters: A review", *Bioresour. Technol.,* vol. 79, no. 3, pp. 251-262, 2001.
[http://dx.doi.org/10.1016/S0960-8524(01)00028-1] [PMID: 11499579]

[9]   C. Park, M. Lee, B. Lee, S.W. Kim, H.A. Chase, J. Lee, and S. Kim, "Biodegradation and biosorption for decolorization of synthetic dyes by Funalia trogii", *Biochem. Eng. J.,* vol. 36, no. 1, pp. 59-65, 2007.
[http://dx.doi.org/10.1016/j.bej.2006.06.007]

[10]   T. Robinson, B. Chandran, and P. Nigam, "Studies on the decolourisation of an artificial textile-effluent by white-rot fungi in N-rich and N-limited media", *Appl. Microbiol. Biotechnol.,* vol. 57, no. 5-6, pp. 810-814, 2001.
[http://dx.doi.org/10.1007/s00253-001-0857-8] [PMID: 11778898]

[11]   S.K. Yesiladalı, G. Pekin, H. Bermek, İ. Arslan-Alaton, D. Orhon, and C. Tamerler, "Bioremediation of textile azo dyes by TrichophytonrubrumLSK-27", *World J. Microbiol. Biotechnol.,* vol. 22, no. 10, pp. 1027-1031, 2006.
[http://dx.doi.org/10.1007/s11274-005-3207-7]

[12]   A.A. Vaidya, and K.V. Date, "Environmental pollution during chemical processing of synthetic fibers", *Colourage,* vol. 14, pp. 3-10, 1982.

[13]   C.G. Boer, L. Obici, C.G.M. Souza, and R.M. Peralta, "Decolorization of synthetic dyes by solid state cultures of Lentinula (Lentinus) edodes producing manganese peroxidase as the main ligninolytic enzyme", *Bioresour. Technol.,* vol. 94, no. 2, pp. 107-112, 2004.
[http://dx.doi.org/10.1016/j.biortech.2003.12.015] [PMID: 15158501]

[14]   R. Maas, and S. Chaudhari, "Adsorption and biological decolourization of azo dye Reactive Red 2 in semicontinuous anaerobic reactors", *Process Biochem.,* vol. 40, no. 2, pp. 699-705, 2005.
[http://dx.doi.org/10.1016/j.procbio.2004.01.038]

[15]   X.C. Jin, G.Q. Liu, Z.H. Xu, and W.Y. Tao, "Decolorization of a dye industry effluent by Aspergillus fumigatus XC6", *Appl. Microbiol. Biotechnol.,* vol. 74, no. 1, pp. 239-243, 2007.
[http://dx.doi.org/10.1007/s00253-006-0658-1] [PMID: 17086413]

[16]   C. O'Neill, F.R. Hawkes, D.L. Hawkes, N.D. Lourenço, H.M. Pinheiro, and W. Delée, "Colour in textile effluents - sources, measurement, discharge consents and simulation: A review", *J. Chem. Technol. Biotechnol.,* vol. 74, no. 11, pp. 1009-1018, 1999.
[http://dx.doi.org/10.1002/(SICI)1097-4660(199911)74:11<1009::AID-JCTB153>3.0.CO;2-N]

[17]   N.K. Kılıç, J.L. Nielsen, M. Yüce, and G. Dönmez, "Characterization of a simple bacterial consortium for effective treatment of wastewaters with reactive dyes and Cr(VI)", *Chemosphere,* vol. 67, no. 4, pp. 826-831, 2007.
[http://dx.doi.org/10.1016/j.chemosphere.2006.08.041] [PMID: 17217991]

[18]   P. Kaushik, and A. Malik, "Fungal dye decolourization: Recent advances and future potential", *Environ. Int.,* vol. 35, no. 1, pp. 127-141, 2009.
[http://dx.doi.org/10.1016/j.envint.2008.05.010] [PMID: 18617266]

[19]   M. Joshi, R. Bansal, and R. Purwar, "Colour removal from textile effluents", *Indian J. Fibre Text. Res.,* vol. 29, no. 2, pp. 239-259, 2004.

[20]   H.S. Freeman, D. Hinks, and J. Esancy, *Physicochemical principles of colour chemistry* A. T. Peters

and H. S. Freeman Editors: Chapman and Hall, Uinted Kingdom, 1996.

[21]   A.B. dos Santos, F.J. Cervantes, and J.B. van Lier, "Review paper on current technologies for decolourisation of textile wastewaters: Perspectives for anaerobic biotechnology", *Bioresour. Technol.*, vol. 98, no. 12, pp. 2369-2385, 2007.
[http://dx.doi.org/10.1016/j.biortech.2006.11.013] [PMID: 17204423]

[22]   W. Logroño, M. Pérez, G. Urquizo, A. Kadier, M. Echeverría, C. Recalde, and G. Rákhely, "Single chamber microbial fuel cell (SCMFC) with a cathodic microalgal biofilm: A preliminary assessment of the generation of bioelectricity and biodegradation of real dye textile wastewater", *Chemosphere*, vol. 176, pp. 378-388, 2017.
[http://dx.doi.org/10.1016/j.chemosphere.2017.02.099] [PMID: 28278426]

[23]   A. Turcanu, and T. Bechtold, "Cathodic decolourisation of reactive dyes in model effluents released from textile dyeing", *J. Clean. Prod.*, vol. 142, no. 4, pp. 1397-1405, 2017.
[http://dx.doi.org/10.1016/j.jclepro.2016.11.167]

[24]   M. Rajasimman, S.V. Babu, and N. Rajamohan, "Biodegradation of textile dyeing industry wastewater using modified anaerobic sequential batch reactor – Start-up, parameter optimization and performance analysis", *J. Taiwan Inst. Chem. Eng.*, vol. 72, pp. 171-181, 2017.
[http://dx.doi.org/10.1016/j.jtice.2017.01.027]

[25]   M.B. Kurade, T.R. Waghmode, S.M. Patil, B.H. Jeon, and S.P. Govindwar, "Monitoring the gradual biodegradation of dyes in a simulated textile effluent and development of a novel triple layered fixed bed reactor using a bacterium-yeast consortium", *Chem. Eng. J.*, vol. 307, pp. 1026-1036, 2017.
[http://dx.doi.org/10.1016/j.cej.2016.09.028]

[26]   A. Pala, and E. Tokat, "Color removal from cotton textile industry wastewater in an activated sludge system with various additives", *Water Res.*, vol. 36, no. 11, pp. 2920-2925, 2002.
[http://dx.doi.org/10.1016/S0043-1354(01)00529-2] [PMID: 12146882]

[27]   S.J. Zhang, M. Yang, Q.X. Yang, Y. Zhang, B.P. Xin, and F. Pan, "Biosorption of reactive dyes by the mycelium pellets of a new isolate of Penicillium oxalicum", *Biotechnol. Lett.*, vol. 25, no. 17, pp. 1479-1482, 2003.
[http://dx.doi.org/10.1023/A:1025036407588] [PMID: 14514054]

[28]   F.P. van der Zee, and S. Villaverde, "Combined anaerobic–aerobic treatment of azo dyes—A short review of bioreactor studies", *Water Res.*, vol. 39, no. 8, pp. 1425-1440, 2005.
[http://dx.doi.org/10.1016/j.watres.2005.03.007] [PMID: 15878014]

[29]   A.A. Dias, R.M. Bezerra, P.M. Lemos, and A. Nazaré Pereira, "*In vivo* and laccase-catalysed decolourization of xenobiotic azo dyes by a basidiomycetous fungus: Characterization of its ligninolytic system", *World J. Microbiol. Biotechnol.*, vol. 19, no. 9, pp. 969-975, 2003.
[http://dx.doi.org/10.1023/B:WIBI.0000007331.94390.5c]

[30]   G. McMullan, C. Meehan, A. Conneely, N. Kirby, T. Robinson, P. Nigam, I.M. Banat, R. Marchant, and W.F. Smyth, "Microbial decolourisation and degradation of textile dyes", *Appl. Microbiol. Biotechnol.*, vol. 56, no. 1-2, pp. 81-87, 2001.
[http://dx.doi.org/10.1007/s002530000587] [PMID: 11499950]

[31]   H. Horitsu, M. Takada, E. Idaka, M. Tomoyeda, and T. Ogawa, "Degradation of p-Aminoazobenzene byBacillus subtilis", *Euro. J. App. Micro.*, vol. 4, no. 3, pp. 217-224, 1977.
[http://dx.doi.org/10.1007/BF01390482]

[32]   E. Idaka, T. Ogawa, H. Horitsu, and M. Tomoyeda, "Degradation of azo compounds by Aeromonashydrophilavar 2413", *J. Soc. Dyers Colour.*, vol. 94, no. 3, pp. 91-94, 1978.
[http://dx.doi.org/10.1111/j.1478-4408.1978.tb03398.x]

[33]   K. Wuhrmann, K. Mechsner, and T. Kappeler, "Investigation on rate? Determining factors in the microbial reduction of azo dyes", *Euro. J. App. Micro. Biotec.*, vol. 9, no. 4, pp. 325-338, 1980.
[http://dx.doi.org/10.1007/BF00508109]

[34]   W. Zhou, and W. Zimmermann, "Decolorization of industrial effluents containing reactive dyes by actinomycetes", *FEMS Microbiol. Lett.,* vol. 107, no. 2-3, pp. 157-161, 1993.
[http://dx.doi.org/10.1111/j.1574-6968.1993.tb06023.x] [PMID: 8472899]

[35]   Z. Aksu, and G. Dönmez, "Combined effects of molasses sucrose and reactive dye on the growth and dye bioaccumulation properties of Candida tropicalis", *Process Biochem.,* vol. 40, no. 7, pp. 2443-2454, 2005.
[http://dx.doi.org/10.1016/j.procbio.2004.09.013]

[36]   J.T. Spadaro, M.H. Gold, and V. Renganathan, "Degradation of azo dyes by the lignin-degrading fungus Phanerochaete chrysosporium", *Appl. Environ. Microbiol.,* vol. 58, no. 8, pp. 2397-2401, 1992.
[http://dx.doi.org/10.1128/aem.58.8.2397-2401.1992] [PMID: 1514787]

[37]   A. Paszczynski, M.B. Pasti-Grigsby, S. Goszczynski, R.L. Crawford, and D.L. Crawford, "Mineralization of sulfonated azo dyes and sulfanilic acid by Phanerochaete chrysosporium and Streptomyces chromofuscus", *Appl. Environ. Microbiol.,* vol. 58, no. 11, pp. 3598-3604, 1992.
[http://dx.doi.org/10.1128/aem.58.11.3598-3604.1992] [PMID: 1482182]

[38]   F. Zhang, J.S. Knapp, and K.N. Tapley, "decolourisation of cotton bleaching effluent with wood rotting fungus", *Water Res.,* vol. 33, no. 4, pp. 919-928, 1999.
[http://dx.doi.org/10.1016/S0043-1354(98)00288-7]

[39]   S. Ohmomo, N. Itoh, Y. Watanabe, Y. Kaneko, Y. Tozawa, and K. Ueda, "Continuous decolorization of molasses wastewater with mycelia of coriolus versicolor Ps4a", *Agric. Biol. Chem.,* vol. 49, no. 9, pp. 2551-2555, 1985.

[40]   J.S. Knapp, P.S. Newby, and L.P. Reece, "Decolorization of dyes by wood-rotting basidiomycete fungi", *Enzyme Microb. Technol.,* vol. 17, no. 7, pp. 664-668, 1995.
[http://dx.doi.org/10.1016/0141-0229(94)00112-5]

[41]   I.K. Kapdan, F. Kargia, G. McMullan, and R. Marchant, "Effect of environmental conditions on biological decolorization of textile dyestuff by C. versicolor", *Enzyme Microb. Technol.,* vol. 26, no. 5-6, pp. 381-387, 2000.
[http://dx.doi.org/10.1016/S0141-0229(99)00168-4] [PMID: 10713211]

[42]   Y. Zeroual, B.S. Kim, C.S. Kim, M. Blaghen, and K.M. Lee, "Biosorption of bromophenol blue from aqueous solutions by Rhizopus stolonifera biomass", *Water Air Soil Pollut.,* vol. 177, no. 1-4, pp. 135-146, 2006.
[http://dx.doi.org/10.1007/s11270-006-9112-3]

[43]   G. Bayramoğlu, and M. Yakup Arıca, "Biosorption of benzidine based textile dyes "Direct Blue 1 and Direct Red 128" using native and heat-treated biomass of Trametes versicolor", *J. Hazard. Mater.,* vol. 143, no. 1-2, pp. 135-143, 2007.
[http://dx.doi.org/10.1016/j.jhazmat.2006.09.002] [PMID: 17010509]

[44]   T. O'Mahony, E. Guibal, and J.M. Tobin, "Reactive dye biosorption by Rhizopus arrhizus biomass", *Enzyme Microb. Technol.,* vol. 31, no. 4, pp. 456-463, 2002.
[http://dx.doi.org/10.1016/S0141-0229(02)00110-2]

[45]   N.S. Maurya, A.K. Mittal, P. Cornel, and E. Rother, "Biosorption of dyes using dead macro fungi: Effect of dye structure, ionic strength and pH", *Bioresour. Technol.,* vol. 97, no. 3, pp. 512-521, 2006.
[http://dx.doi.org/10.1016/j.biortech.2005.02.045] [PMID: 16216733]

[46]   H. Ali, "Biodegradation of synthetic dyes – A review", *Water Air Soil Pollut.,* vol. 213, no. 1-4, pp. 251-273, 2010.
[http://dx.doi.org/10.1007/s11270-010-0382-4]

[47]   S. Senthilkumar, M. Perumalsamy, and H. Janardhana Prabhu, "Decolourization potential of white-rot fungus Phanerochaete chrysosporium on synthetic dye bath effluent containing Amido black 10B", *J. Saudi Chem. Soc.,* vol. 18, no. 6, pp. 845-853, 2014.
[http://dx.doi.org/10.1016/j.jscs.2011.10.010]

[48]   G. Dönmez, and Z. Aksu, "Removal of chromium(VI) from saline wastewaters by Dunaliella species", *Process Biochem.,* vol. 38, no. 5, pp. 751-762, 2002.
[http://dx.doi.org/10.1016/S0032-9592(02)00204-2]

[49]   S.A. Misal, D.P. Lingojwar, R.M. Shinde, and K.R. Gawai, "Purification and characterization of azoreductase from alkaliphilic strain Bacillus badius", *Process Biochem.,* vol. 46, no. 6, pp. 1264-1269, 2011.
[http://dx.doi.org/10.1016/j.procbio.2011.02.013]

[50]   V. Ferreiraleitao, M. Decarvalho, and E. Bon, "Lignin peroxidase efficiency for methylene blue decolouration: Comparison to reported methods", *Dyes Pigments,* vol. 74, no. 1, pp. 230-236, 2007.
[http://dx.doi.org/10.1016/j.dyepig.2006.02.002]

[51]   C. Rani, A.K. Jana, and A. Bansal, "Studies on the biodegradation of azo dyes by white rot fungi Phlebia Radiata", *Proceedings of International Conference on Energy and Environment*, 2009, pp. 203-207.

[52]   Tongbu Lu, Xinyu Peng, Huiying Yang, and Liangnian Ji, "The production of glucose oxidase using the waste myceliums of Aspergillus niger and the effects of metal ions on the activity of glucose oxidase", *Enzyme Microb. Technol.,* vol. 19, no. 5, pp. 339-342, 1996.
[http://dx.doi.org/10.1016/S0141-0229(96)00004-X]

[53]   M.P. Miranda, G.G. Benito, N.S. Cristobal, and C.H. Nieto, "Color elimination from molasses wastewater by Aspergillus niger", *Bioresour. Technol.,* vol. 57, no. 3, pp. 229-235, 1996.
[http://dx.doi.org/10.1016/S0960-8524(96)00048-X]

[54]   A. Al-Kdasi, A. Idris, K. Saed, and C.T. Guan, "Treatment of textile wastewater by advanced oxidation processes: A review", *Global Nest: Int. J.,* vol. 6, no. 3, pp. 222-230, 2004.

[55]   L. Coulibaly, G. Gourene, and N.S. Agathos, "Utilization of fungi for biotreatment of raw wastewaters", *Afr. J. Biotechnol.,* vol. 2, no. 12, pp. 620-630, 2003.
[http://dx.doi.org/10.5897/AJB2003.000-1116]

SUBJECT INDEX

www.ingramcontent.com/pod-product-compliance
Lightning Source LLC
Chambersburg PA
CBHW041659210326
41598CB00007B/463